MATH GAMES & ACTIVITIES

Vol. 1, A Toolbox of Duplicating Designs for Early and Middle Grade Elementary School Teachers

Paul Shoecraft

Math Games and Activities Vol. 1
ISBN 0-86651-330-2
(previously ISBN 0-941530-01-9)
DS01721

Math Games and Activities Vol. 1 and Vol. 2 Set
ISBN 0-86651-332-9
(previously ISBN 0-941530-00-0)
DS01723

9 10 -ML- 99 98 97

DALE
SEYMOUR
PUBLICATIONS
P.O. BOX 10888
PALO ALTO, CA 94303

Acknowledgments

For much of the contents of this book I am in debt to other people, some of whose names I have forgotten, some of whose names are mentioned in the bibliography, and some of whose names are given below.

Students:

Shaun Hill of Churchlands College of Advanced Education in Western Australia for the spark which led to **Super Bowl**. Marianne Addison, Julie Advic, Brian Argus, Robyn Bevis, Shaunna Bruce, Chris Catchpole, Karen Cocking, Lynn Dickens, Karen Hayes, Isabel Laurance, Sharon McDermott, Janet McLay, Anne Playfield, and Wendy Stonehouse, all of Churchlands College of Advanced Education, for their suggestions for basic facts activities.

Teachers:

Jonathan Knaupp of Arizona State University for many of the colored rod games. Julie Martin of the Western Australian schools for the many ideas for the **Numeral cards**. And Thomas Palumbo of the Philadelphia schools for **One Less** and **Something First**.

Publishers:

Addison-Wesley Publishing Company, Inc. for permission to reprint the **Trizo gameboard** and the small multi-base blocks from the instructor's manual to **BASIC MATHEMATICS: A BLUEPRINT FOR SUCCESS**. The National Council of Teachers of Mathematics for permission to reprint **Me as a Measure** and **Merry Measuring** from **THE ARITHMETIC TEACHER**. The Mathematical Association of Western Australia for permission to reprint **Ollie Octopus** from **RHOMBUS**. And the Arizona Silver Belt for the use of their typesetting equipment.

Artists:

Joy Scott of Churchlands College of Advanced Education for her contributions to the non-technical art. Mike Quaintance of Bailey, Colorado for the cover photograph.

To my wife, Lynne, my one true love, who ignored the fact that I had to ignore her more than I wanted to to write this.

TABLE
OF
CONTENTS

RESOURCE PAGE DIRECTORY

Resource	Location		Description					
	Directions	Duplicating design(s)	Manipulative	Game	Activity	Worksheet	Instructional aid	General Supplies
Action Cards	01	79-80					●	
Add-a-Man	01	139				●		
Addition and Subtraction Facts Dominoes	01	167-172		●				
Addition Facts Rummy	02	146-152		●				
Addition Facts Table	04	128					●	
Addition in Different Lands	06	103-104			●			
Applingo	07	191-198		●				
ASMD Cards	08	251-254					●	
ASMD Scorecard	010	256					●	
Attribute Shapes	010	14-15	●					
Balance Beam Cutout	010	129	●					
Balance Beam Worksheets	014	130-132			●			

Resource	Location		Description					
	Directions	Dupli-cating design(s)	Manip-ulative	Game	Acti-vity	Work-sheet	Instruc-tional aid	General Supplies
Balancing Act	015	52-53		●				
Balloon Pop	016	155-156		●				
Bank It or Clear It	016	99-100		●				
Blank Cards	018	6-8						●
Blank Dominoes	019	10						●
Blank Tags	019	9						●
Block It or Shade In	020	38-39	●					
Boat	021	141	●					
Bridge	022	46-47	●					
Bubbles	024	74				●		
Build a Cube or Break a Cube	024	101	●					
Caboose	025	188	●					
Car Park	025	25-26	●					
Circus	026	206-207				●		
Clown	026	54-55	●					
Collect-a-Shape	027	59-60	●					
Color the Shapes	027	278				●		
Computasnake	027	164-165				●		
Connect-a-Dot	027	87-90				●		
Countdown	028	220				●		
Count the Shapes	028	279				●		
Crossnumber Puzzles	028	183-184, 214				●		

Resource	Location		Description					
	Directions	Duplicating design(s)	Manipulative	Game	Activity	Worksheet	Instructional aid	General Supplies
Digit Cards	029	120-121					●	
Dinosaur Eggs	029	162-163			●			
Division Cards	029	210-211					●	
Division Worksheet	029	231				●		
Doll	030	236-240		●				
Doll House	031	242-246		●				
Dot Paper	031	264-265					●	
Double Nine Dominoes	035	32-37		●				
Engine	038	143		●				
Experience Roster	038	1						●
Find the Cheese	039	300		●				
Fish	039	110				●		
Fly, Fly Away	039	227-228				●		
Fraction Dominoes	039	272-274		●				
Fraction Rulers	040	269-270					●	
Frog	040	133-134				●		
Geo Dominoes	040	280-282		●				
Gimmi	041	16, 21		●				
Going to the Park	041	12-13		●				
Grid Paper	042	313-314					●	
Hex	042	50		●				
Hiking	043	118-119		●				

Resource Page Directory, Continued

Resource	Location		Description					
	Directions	Duplicating design(s)	Manipulative	Game	Activity	Worksheet	Instructional aid	General Supplies
Horse	043	142		●				
House	044	186		●				
Hundred Chart	044	178		●			●	
Improve Your Aim	045	225-226				●		
Jellybeans	045	109				●		
Kilometer Count	045	138				●		
Lady Bugs	045	159-160				●		
Lattice Multiplication	045	230				●		
Leaf Patterns	047	176-177				●		
Lily Pad	047	48-49		●				
Line Designs	048	296-299			●			
Liter Cutout	050	325-326			●			
Man	050	185		●				
Me as a Measure	050	301			●			
Merry Measuring	050	271			●			
Metric Concentration	050	311-312		●				
Milliliter, 10-Milliliter, and 100-Milliliter Cutouts	051	324			●			
Mirror Symmetry	051	294-295			●			
Monster	053	137				●		
Moon Walk	053	223		●				
Motley and Mates	054	263					●	
Motley Crab Adder	054	257					●	

Resource	Location		Description					
	Directions	Duplicating design(s)	Manipulative	Game	Activity	Worksheet	Instructional aid	General Supplies
Multi-base Blocks	058	93-98	●					
Multiplication Facts Rummy	059	199-205		●				
Multiplication Facts Table	059	175					●	
Non-digital Clocks	060	275-277					●	
Number Houses	060	161				●		
Numeral Cards	060	75-78		●			●	
Numeral Dominoes	061	81-86		●				
Numeral Puzzles	062	67-72			●			
Numeration in Different Lands	062	102				●		
Ollie Octopus	063	122-127		●				
One Less	064	174				●		
One More / Less	064	91-92				●		
Optical Illusions	064	304			●			
Paths	064	303			●			
People Pictures	065	11	●					
Pick-a-Pair	065	29-31		●				
Place Value Rummy	065	111-117		●				
Puppies	066	135-136				●		
Racer	066	221-222				●		
Ring Around the Rosy	066	189-190		●				
Rod Patterns	066	44-45			●			

Resource Page Directory, Continued

Resource	Location		Description					
	Directions	Duplicating design(s)	Manipulative	Game	Activity	Worksheet	Instructional aid	General Supplies
Rulers	067	302						●
Safari	067	56-58		●				
Sameness Trains	067	17, 22			●			
Seesaw	068	157-158				●		
Sequence Cards	068	40-43			●			
Shape Rummy	068	283-289		●				
Shooting Gallery	069	144-145		●				
Shopping Cards	069	266-268					●	
Sir Crab Multiplier	070	259					●	
Ski Slope	071	208-209		●				
Smoke Rings	072	73				●		
Snail Trail	072	166		●				
Something First	073	173				●		
Speedy Operator	073	232-233				●		
Spider and Fly	074	234-235		●				
Spinners	075	2-5		●				●
Staircase	076	140		●				
Stepping Stones	076	216-219		●				
Subtraction in Different Lands	076	105-106			●			
Subtraction Magic	077	229				●		
Super Bowl	078	249-250		●				
Take the Children Home	079	179-180				●		

Resource	Location		Description					
	Directions	Duplicating design(s)	Manipulative	Game	Activity	Worksheet	Instructional aid	General Supplies
Tangram Pieces	079	315	●					
Tangram Task Cards	080	316-323			●			
Teeter Totter	082	212-213				●		
Telephone	082	62-64		●				
Ten Land Task Cards	082	107-108				●		
The Great Legalizer	083	261					●	
The Impeccable Twin Dividers	084	260					●	
The Magnificent Equalizer	086	262					●	
The Scruffy Twin Subtractors	087	258					●	
Three-Way Sort Task Card	091	24					●	
Toys	092	153-154				●		
Tree	092	187		●				
Trizo	092	224		●				
Two-Way Sort Task Card	093	23					●	
Vertices, Faces, and Edges	095	291-293				●		
Walk a Crooked Meter	096	305		●				
Watermelon	097	27-28		●				
Witch	098	181-182				●		
Wobble Town	098	290			●			
Word Cards	0101	18-20	●					

Resource	Location		Description					
	Direc-tions	Dupli-cating design(s)	Manip-ulative	Game	Acti-vity	Work-sheet	Instruc-tional aid	General Supplie
World Series	0101	247-248		●				
Writing Numerals	0102	65-66				●		
Ziggy's Home Run	0102	306-310		●				

INTRODUCTION

Description

MATH GAMES & ACTIVITIES is a two-volume resource for teaching mathematics in the elementary school. The first volume is for grades K-5, the second for grades 4-9. In that this is the first volume, what is said in what follows is about this volume in particular.

MATH GAMES & ACTIVITIES is essentially two things: One, it is a resource for teachers who like to teach, who like to say to children things like "Do this or that," "Look at this or that," "Think about this or that," who like to say these sorts of things more than things like "Turn to page blah-blah-blah and do the 'odd' exercises." In this regard, it contains nearly 150 "ideas" on how to enhance one's teaching of mathematics. Two, it is a resource which attempts to account for the fact that however easily an idea might be communicated, and however much a teacher might want to implement it, that to implement it typically requires the production of materials, a time consuming and energy draining enterprise, particularly in relation to being responsible for 20 or more children in one's professional life and being responsive to friends and family in one's personal life. Thus in addition to the ideas, it contains more than 300 duplicating designs with which to easily and inexpensively implement the ideas.

Organization

To obtain an overview of the resource provided by MATH GAMES & ACTIVITIES, begin by skimming through the Resource Page Directory beginning on page vii. In doing so, note that the ideas listed there are listed in alphabetical order in both the directory and in the front matter immediately following the directory, that the ideas are categorized in terms of

- ✔ manipulatives

- ✔ games

- ✔ activities

- ✔ worksheets

- ✓ instructional aids

- ✓ general supplies

and that the ideas are located in the front matter and among the duplicating designs by two sets of page numbers, the first having to do with the whereabouts of the directions for the ideas, the second with the whereabouts of the duplicating designs for the ideas. (The page numbers for the directions are preceded with a zero to distinguish them from the page numbers for the duplicating designs.)

Then pick a topic from the **Table of Contents** and "browse" through the duplicating designs the topic directs you to. As you do so, think about how the duplicating designs are probably used. Then turn to the **Resource Page Directory** again and locate the page numbers of the directions for the duplicating designs and read some of the directions. Chances are that what you read will be very much like what you thought you would read.

Finally, turn to the **Index** beginning on page 327 and note the way it categorizes the duplicating designs and directs you to other material in **MATH GAMES & ACTIVITIES** which touches on the same topic.

Throughout, note the shading and the guides along the edges of some of the pages of the book to facilitate leafing through the book.

Use

In using **MATH GAMES & ACTIVITIES**, the key consideration is always what you want to make from the duplicating designs in it. This determines the sort of equipment you will need to bring to bear on them.

If you want to make transparencies of them, simply adjoin them one at a time to a piece of transparency film and run them and the film through a Thermofax machine.

If you want to make single copies of them, simply Xerox them.

If you want to make single copies of them, but you want the copies colored, backed, and / or laminated, Xerox them and see to the coloring, backing, and / or laminating of the copies as you perhaps watch your favorite television program.

If you want to make multiple copies of them, make ditto masters for them using a Thermofax machine and make the copies from the ditto masters using a ditto machine.

If you want to make multiple copies of them, but you want the copies colored, backed, and / or laminated, make the copies as already mentioned and have your class or a parent group help with the coloring, backing, and / or laminating of them. If the children in your class are too young to help with this, see about "borrowing" some older children from another class.

If you want to make multiple copies of them on poster board so as to not have to back them, cut stencils for them using a Thermofax machine and print from the stencils directly onto poster board using a Gestetner duplicator. (Some schools have reprographic centers which will do this for you and will even "cut along the solid or dotted lines" for you.)

Note that all that is ever done with the duplicating designs is copy them. Thus they should last indefinitely for any number of different uses.

Another thing to consider in using **MATH GAMES & ACTIVITIES** is the cost of the supplies involved, the paper, poster board*, paste, crayons, transparency film, and laminating film. Note, however, that the cost is easily reconciled by comparing it to what really costs in education, the buildings, buses, utilities, and most of all, the services of maintenance and custodial persons, teachers, counselors, and administrators, none of which or whom, in themselves, teach anyone anything. Thus the thing to keep in mind if worried about the cost of supplies is that they are part of the "delivery" system, that they, along with what we say to children and have them do or read, are what all the buildings, buses, utilities . . . are there for, to have children learn. This is not to mean that we should be wasteful of supplies, but rather that we should never be so watchful of them to where we reduce our effectiveness in the classroom to where what is spent on buildings, buses, utilities . . . is largely wasted.

A cost-free alternative to poster board for backing purposes is empty cereal boxes, a resource children can supply in abundance.

A man unusually committed to life in all things observed a hill of ants in a farmer's field. Looking up, he saw the farmer headed toward the ants with his plow. The man thought, "If only I were an ant, I could tell the ants of their peril that they might move out of the way of the farmer's plow." So he became an ant.

CONTENTS

Manipulatives

MATH GAMES & ACTIVITIES contains duplicating designs for the following manipulative materials:

- ✔ attribute shapes

- ✔ balance beams

- ✔ multi-base blocks

- ✔ people pictures

- ✔ tangram pieces

- ✔ word cards

In addition, it contains games, worksheets, and task cards with which to direct the use of the manipulative materials as given under the headings in the front matter for the manipulative materials.

If your classroom is not equipped with these manipulative materials, or is not equipped with them in sufficient amounts, you can make reasonable facsimilies of them from the duplicating designs for such. To this end, you would make suitable numbers of copies of the duplicating designs and have your class or a parent group color and back the copies and afterwards cut them out. If you have them laminate them before cutting them out, the manipulative materials you end up with will be nearly as attractive and long lasting as those made of wood or plastic and, with the possible exception of the multi-base blocks, will certainly be as educational.

If your classroom is already equipped with sufficient quantities of the above mentioned manipulative materials, then, instead of making more of them, you could make copies of the duplicating designs for them, and with the copies make task cards or flannel board materials with which to direct the use of the manipulative materials. You could also make transparencies of them, and with the transparencies make materials for directing the use of the manipulative materials with an overhead projector.

As to why we teach with manipulative materials, the reasons are many and varied:

They hold children's interest.

They allow for the development of a vocabulary common to both teacher and children.

They provide for individual differences with respect to ways of approaching a problem.

They permit verification of results.

But the most important reason for teaching with manipulative materials is that they give meaning to mathematics in that they are the "something else" by which it is understood. To see this, note that nothing can be understood solely in terms of itself. To illustrate, if you were to look up **ancespitorian** in the dictionary and were to find the following, you still would not know what it meant.

an·ces·pi·tor·i·an (ăn-sĕs′pĭ-tôr′ē-ən) **adj.** what is said about something which is ancespitorian

At best, all you would know about **ancespitorian** is how to spell and pronounce it. To know what it meant, you would have to know how it was like or unlike something else which you already knew about.

Incidentally, if you try to find **ancespitorian** in a dictionary, you will not find it. It is just a made-up word with which to make a point. Still, if you wanted to make it into a word, a good definition for it would be "anything which can be 'spelled' and pronounced but which is not understood." With this definition, one could then say that the objective of teaching with manipulative materials is to achieve in children's minds more than a mere ancespitorian knowledge of mathematics.

Games

MATH GAMES & ACTIVITIES contains instructions and duplicating designs for more than 50 games -- card games, dominoe games, board games bingo games, and other types of games, the titles and locations of which are given in the **Resource Page Directory**. Some of them, like **Bank It or Clear It** are variations or refinements of games you might already know. But some o them, notably, **Doll, Doll House, World Series,** and **Super Bowl,** are new to this publication. If copies of the duplicating designs for the games are colored backed, and laminated, the result is educational materials with striking appea and marked durability.

In viewing the games, note that nearly all of them can be adapted to a variety of topics. For example, **Watermelon,** a game on making one-to-one correspondences, is easily made into a game on the basic facts by modeling it after **Applingo** and writing on the "seeds" and in the seed spaces on the gameboard for the game. And **Clown,** a counting game, is likewise easily made into a game on the basic facts by modeling it after **Stepping Stones** and writing in the circles on the gameboard for the game. In addition, some of the gameboards come in two forms, the one being on a particular topic and ready for use, the other being blank as far as a topic is concerned but ready for altering. If you ask yourself how you might adapt this or that game to a topic different from the one it is for, the answer will typically jump into your head.

As to the purpose of the games in **MATH GAMES & ACTIVITIES,** they are used primarily to focus attention on key concepts such as base and place value and to motivate drill and practice on things like the basic facts which need to be committed to memory. In this regard, the games can be used to stock a math lab or an interest center, in which case only one or two copies of the games wanted would be needed. However, a more exciting use of the games is to use them to actually teach with, in which case 10 or more copies of the games wanted would be needed, enough for an entire class. To experiment with the feasibility of teaching with games, begin with a game like **Target Practice** with a consumable gameboard. In this way your investment in the experiment in terms of supplies used and time spent will be negligible.

Some things to consider in using the games in **MATH GAMES & ACTIVITIES** are the following:

Before reading the directions for a game, have a good look at the duplicating designs for the game. In this way you will probably anticipate many of the directions for the game and will therefore be reading the directions more for confirmation than for understanding.

Note that there is no right or wrong way to play the games. Rather, there are effective ways to play them, ways in which to meet the educational objectives of the games. Thus the directions for the games can be altered to focus on other objectives or to make the games easier or harder for certain children.

Note that the directions for the board games with "finish" spaces on the gameboards do not specify how one gets to the finish spaces. Does one just have to get there? Or does one have to get there exactly? Thus a ruling needs to be made on this. One such rule is the "bounce back" rule, a rule that if a player moves into a finish space but still has some moves to go, the player must "bounce back" the number of moves to go. To illustrate, if a player were three moves away from the finish space and rolled a 5, the player would move 1, 2, 3 and into the finish space, and then 4, 5 and right back to where the player started from.

To facilitate the making of classroom quantities of a game, make multiple copies of the game on poster board as explained under **Use.** In this way you can even save on the coloring in by choosing a suitable color of poster board.

When making copies of the duplicating designs for a game, always make a few more copies than you intend to make up and store them in a folder. In this way you can easily replace a game which gets lost or destroyed.

To make one of the gameboards where the gameboard is in two pieces, lay next to one another the edges of the gameboard which are to be joined and tape along the edges with a piece of clear tape as long as the gameboard is wide. (See Figure 1.)

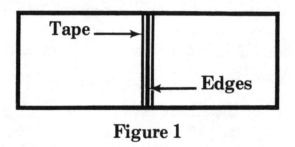

Figure 1

To strengthen the join on the gameboard, use two pieces of tape, one piece on top of the other. Note: Do **not** tape on both sides of the gameboard. If you do, the gameboard will not fold easily. Thus if you want the gameboard to fold in (so as to protect the playing surface of the gameboard while the gameboard is in storage), tape on the playing surface of the gameboard. If you want the gameboard to fold out (so as to draw attention to the gameboard while the gameboard is on display as in a clear plastic bag), tape on the back of the gameboard.

Keep the time which you personally invest in the making of games to a minimum. In this way, you won't mind so much if some of them get lost or scruffed up. What is wanted are games which are valued but not to the point of reluctance at letting children play them and perhaps take them home with them.

Always store the games flat rather than rolled. If you roll them, you can hardly ever get them to lie really flat again.

If making games for children who can read, make copies of the directions for the games as well and mount the copies on poster board or on the games themselves that they might be included with the games. (Note that the directions for the games are in boldface type to make them easy to read if copied.)

Make transparencies of the duplicating designs for a game that the rules for the game might be easily explained to an entire class with an overhead projector.

As a follow-up to the card games and dominoe games, use transparencies of the cards and dominoes with an overhead projector as flash cards for a class.

Involve children as much as possible in the making of the games. In doing so, you not only end up with lots of games, but with children who have a personal investment in them and therefore an interest in playing them.

As to why one should consider teaching with games, note the following: In addition to simulating real world uses of mathematics and making pleasurable the drill and practice necessary for mastery, they

- ✔ motivate

- ✔ build confidence

- ✔ improve attitudes

- ✔ encourage peer teaching

- ✔ allow for individual differences

The latter point is most apparent with games, because the deciding factor in who wins in games is typically luck, not ability. For this reason, games can even be used in tutoring situations without diminishing the interest value of the games, because the one being tutored is just as likely to win as the one doing the tutoring.

Activities

Activities refer to many things in education: children manipulating objects, working busily at their desks, or moving about the classroom or outdoors. The main ingredient is involvement -- involvement of the children in what they are doing. Another ingredient is the direction, either written or spoken, the children are given. Examples from **MATH GAMES & ACTIVITIES** of direction which leads to involvement would be **Dinosaur Eggs**, a puzzle on the basic facts, **Merry Measuring**, an exercise on measuring with the body, and **Wobble Town**, a construction activity with straws and pipe cleaners.

Worksheets

The worksheets in **MATH GAMES & ACTIVITIES** are on a variety of topics: cardination, ordination, numeration, the number facts, equality, and others. They appear in forms known to be appealing to children: connect-a-dots, coloring exercises, counting tasks, puzzles, crossnumber puzzles, and others. They are distinguished from the activities in **MATH GAMES & ACTIVITIES** in that they are self-contained.

A reason for using the worksheets in **MATH GAMES & ACTIVITIES** is to obtain a written record of what children know, a record which can be used for evaluation purposes. If duplicated as transparencies or in classroom quantities, they can be used to direct a lesson or series of lessons. They can also be used to stock an interest center or to make individual assignments.

Instructional Aids

In general, an instructional aid differs from an activity or a worksheet in that an instructional aid is something a teacher uses instead of something children use. It is used to get the attention of children or to give children something they can react to in response to teacher directives.

The instructional aids in **MATH GAMES & ACTIVITIES** are of the sort a teacher would use in presenting a lesson to an entire class or in diagnosing strengths and weaknesses in whole number arithmetic. An example of the former would be a lesson on numeration with the **Action cards** and **Numeral cards**. An example of the latter the use of the **ASMD cards** and the **ASMD scorecard** in conjunction with the games of **Doll**, **Doll House**, **World Series**, and **Super Bowl**.

General Supplies

The general supplies in **MATH GAMES & ACTIVITIES** consist of things like rulers and spinners which have use in general, the **Experience Roster** which is for record keeping, and blank cards, dominoes, and tags which can be used to add to the games and activities in **MATH GAMES & ACTIVITIES**. Their location is given in the **Resource Page Directory.**

DIRECTIONS IN ALPHABETICAL ORDER

Action Cards (pp. 79-80)

Sixteen actions which can be performed with the body.

The **Action cards** are used in conjunction with the **Numeral cards** on pages 75-78 to motivate drill and practice on the numerals 1, 2,.3, . . . , 12. The procedure is for a teacher, or a "group leader" amongst a group of children, to hold up an **Action card** and a **Numeral card** as a directive for an audience to do whatever the **Action card** says the number of times the **Numeral card** says. The rule is that if the person holding the cards laughs, the person must relinquish the cards and let someone else hold up the cards.

Add-a-Man (p. 139)

A worksheet on simple sums.

The man "adds up" to 43, that is, is 43 years old. A variation on **Add-a-Man** is to have children draw some number people of their own to give to one another to add up.

Addition and Subtraction Facts Dominoes

A game for two to four on the addition and subtraction facts.

Materials:

One set of Addition and Subtraction Facts Dominoes (pp. 167-172) per group.

Directions:

The same as for the amateur version of Double Nine Dominoes.

Some examples of some matches for the Addition and Subtraction Facts Dominoes would be the "1" and "1 + 0," the "two" and "6 — 4," and the "4 + 5" and "15 — 6." Note: The Addition and Subtraction Facts Dominoes are compatible with the Double Nine Dominoes and the Numeral Dominoes.

Addition Facts Rummy

A game for two to four on the addition facts.

Materials:

One deck of Addition Facts Rummy cards (pp. 146-152) per group.

Directions:

Addition Facts Rummy is played the same as regular rummy.

To begin, each player draws a card. The player drawing the highest card deals, and the player to the left of the person dealing plays first. The play rotates clockwise. The person dealing shuffles the deck, deals seven cards to each player, lays what is left of the deck face down, and turns the top card of the deck face up and lays it next to the deck to start the discard pile.

To play, a player takes either the top card of the deck or the top card of the discard pile. The object is to make "spreads," "books," and "runs." A spread is three of a kind such as the "1 + 3," "2 + 2," and "4." A book is four of a kind such as the "3 + 1" as well. And a run is a sequence of three or more cards that increases by one such as the "3 + 2," "6," and "2 + 5." The player then discards one card, and the play goes to the next player.

When a player makes a spread, book, or run, the player lays it face up. Also, when a player draws the fourth card to another player's spread or draws a card that would play on another player's run, the player may, at his or her option, lay the card face up.

The play stops when a player lays all of his or her cards face up. Each player then gets five points for each card face up minus five points for each card still in his or her hand.

The first player to get 100 points wins.

Note: A rummy deck can also be used to play other card games, notably, Fish, War, and Concentration. The rules for each of these games follow:

FISH

To begin, each player draws a card. The player drawing the highest card deals, and the player to the left of the person dealing plays first. The play rotates clockwise. The person dealing shuffles the deck, deals five cards to each player, and lays what is left of the deck face down.

To play, a player, say, player A, looks at his or her cards and asks any other player for a card like one of them. If the other player has the card, he or she must give it to player A who lays it face up with the card it is like to make a "pair." (An example of a pair for the Addition Facts Rummy cards would be the "1 + 3" and "2 + 2.") Player A then repeats the process until he or she asks for a card that another player does not have, in which case the player tells player A to "fish." Player A then draws the top card from the deck, and the play goes to the next player.

If a player runs out of cards before the deck is exhausted, the player draws the top three cards from the deck and resumes play as his or her turn would dictate.

The player with the most cards after all the cards have been paired wins.

WAR

To begin, each player draws a card. The player drawing the highest card deals, and the player to the left of the person dealing plays first. The play rotates clockwise. The person dealing shuffles the deck and deals out ALL the cards to the players as the players stack their cards without looking at them.

To play, the players take turns putting the top cards from their stacks of cards face up on top of one another. Then, if a card so placed would make a pair with the card directly beneath it, the players race to put a hand on the pile thus formed. The first player to do so gets all the cards in the pile.

The player who ends up with everyone else's cards wins.

CONCENTRATION

To begin, the players select any eight pairs from the Addition Facts Rummy cards to make a "short" deck. The players then draw one card each from the short deck. The player drawing the highest card shuffles the short deck and spreads it out face down to make a four-by-four array. The player to the left of the player making the array plays first. The play rotates clockwise.

To play, a player draws two cards from the array. If the cards make a pair, the player keeps the cards and continues to draw cards two at a time until he or she draws two cards which do not make a pair, in which case he or she returns the cards which do not make a pair to the array exactly as they were, and the play goes to the next player.

The player with the most pairs after all the cards have been drawn wins.

Addition Facts Table (p. 128)

A list of all 100 addition facts.

The most fundamental use of the **Addition Facts table** is as a referent for the "addition facts algorithm," a relatively new procedure for adding, which, for lengthy problems, has certain advantages over the conventional procedure for adding. An example of the algorithm in use is given below.

$$
\begin{array}{cccc}
 & \textcircled{1} & \textcircled{2} & \\
 & {}_1 9_0 & 0_2 & 8 \\
 & 8_8 & 7_9 & {}_1 6_4 \\
+ & {}_1 3_1 & {}_1 6_5 & {}_1 9_3 \\
\hline
2 & 1 & 5 & 3
\end{array}
$$

In the one's column, the 8 was added to the 6, and the sum, or 14, was written on either side of the 6. Then the 4 from the 14 was added to the 9, and the sum, or 13, was written on either side of the 9. Then the 3 from the 13 was written in the one's column of the sum, and the two ones (or tens, actually) to the left of the 6 and 9 were "carried" to the top of the ten's column. And so on.

The most striking feature of the algorithm is that it never has one adding more than two single-digit numbers at a time, thus it never taxes a user beyond the limits of the **Addition Facts table.** Thus the algorithm can be used by even first and second graders that they might work with more than two addends and see for themselves that there is more to adding than "carrying a one," a likely conjecture if all they ever work with are two addends. In addition, the algorithm is a direct reflection of addition in terms of multi-base blocks (the ones to the left of a column indicate that an exchange has been made, the digits to the right of the column that there were that many of whatever left over), is non-stressful in that it allows a user to take a break anywhere in a problem and later on pick up where he or she left off, and is diagnostic in that it allows a teacher to see where in a problem things went wrong in the event of a wrong answer. Its only drawbacks are that it requires a lot of writing, which takes getting used to, and, with all the writing, can get messy and therefore confusing. However, the need for all the writing is offset to some extent with a tendency toward greater accuracy, and the problem of messiness can be dealt with, in part, by having users spread out the digits in a problem in writing the problem.

An additional use of the **Addition Facts table** is to relate addition and multiplication by having children discover the rule for finding the sum of all 100 addition facts. In preparation, make either a transparency of the table for overhead projection or one copy of the table per child for worksheet purposes. Then present the table in conjunction with the following "clues":

> "Circle any three horizontally adjacent numbers. What's an easy way to find their sum?" (Multiply the middle number by three -- the number of numbers.)

> "Circle any three vertically adjacent numbers. What's an easy way to find their sum?" (Multiply the middle number by three -- the number of numbers.)

> "Circle any 'block of four' (any two-by-two array of numbers). What's an easy way to find their sum?" (Multiply the repeating number by four -- the number of numbers.)

> "Circle any 'cross of five' (three horizontally adjacent numbers intersecting three vertically adjacent numbers in the middle). What's an easy way to find their sum?" (Multiply the middle number by five -- the number of numbers.)

> "Circle any 'block of nine' (any three-by-three array of numbers). What's an easy way to find their sum?" (Multiply the number that repeats the most by nine -- the number of numbers.)

> "How would you find the sum of all 100 numbers in the table?" (Multiply nine, the number that repeats the most, by 100 -- the number of numbers.)

For additional discoveries related to the table, have children discover the rules for adding any 2, 3, 4 and so on diagonally adjacent numbers or any 2, 3, 4 and so on columns of numbers where the columns are 2, 3, 4 and so on rows high and skip a column of numbers each time. Examples of three diagonally adjacent numbers and three columns of numbers where the columns are three rows high and skip a column each time are given in Figures 1 and 2, respectively.

6	-	-
-	8	-
-	-	10

Figure 1. The sum is three times the middle number.

8	- 10	- 12
9	- 11	- 13
10	- 12	- 14

Figure 2. The sum is nine times the middle number.

Addition in Different Lands

An activity on the concepts of base and place value in relation to addition of whole numbers.

The objective of **Addition in Different Lands** is not to teach children how to add in different bases, but rather when and how to exchange when adding. To this end, it illustrates that base is only a number mandating an exchange and place value only a convention for recording that an exchange has been made. This is why it involves different bases: to strengthen the concept of exchanging by varying on the number mandating an exchange. This is also why it is couched in "lands" terminology -- the "going to jail" stuff below -- to give children a playful reason for exchanging. The overall aim of the activity is for children who, when told to add in ten land (just like their moms and dads), will properly exchange ten of one kind for one of the next larger kind.

Materials:

For each child, a set of multi-base blocks or a four-color assortment of counters and one of either of the two **Addition in Different Lands worksheets** (pp. 103-104).

Procedure:

For best results, precede with **Bank It** and **Build a Cube**. Then present the worksheet in a way that covers the following points:

For each problem, note the land it is in and illustrate

the addends (the numbers or amounts to be added) with the blocks (or counter equivalent of the blocks) for that land by illustrating a numeral in the unit column with that many units, a numeral in the long column with that many longs, a numeral in the flat column with that many flats, and a numeral in the cube column with that many cubes.

Once the addends are illustrated with the blocks, combine the blocks and, to avoid breaking the law of the land and having to go to jail (as explained under **Bank It**), take the too many of whatever to the "bank" and exchange them for wood equivalents.

What you end up with will be the sum. Record it beneath the addends being careful to record the digits for it in the proper columns.

The answers for the worksheets are as follows:

Page 103:

a. 1001
b. 1100
c. 1020
d. 2210
e. 3215

Page 104:

a. 2012
b. 6430
c. 4071
d. 5373
e. 10313

Applingo

A game for two to four on the multiplication facts. With different numbers, as with the blank Applingo apples on page 198 and the blank Applingo card on page 195, a game on any of the basic facts.

Materials:

One set of Applingo apples (pp. 196-197) per group. One Applingo card (pp. 91-194) per player.

Directions:

To begin, each player "picks" an apple. The player picking the apple with the greatest product is the "group leader." All the apples are then laid face down and scrambled.

To play, the group leader turns up an apple as each player, including the group leader, tries to match the apple with an apple space on his or her card. If a match can be made, the player making the match puts the apple on the apple space on his or her card, and the group leader turns up another apple as before. If no match can be made, the group leader keeps turning up apples until a match can be made.

The first player to fill in all the apple spaces on his or her card wins.

ASMD Cards (pp. 251-254)

More than 125 problems on whole number arithmetic. (ASMD is an acronym for addition, subtraction, multiplication, and division.) The key for the problems is on page 255.

The ASMD cards are used with Doll, Doll House, World Series, and Super Bowl to diagnose weaknesses in whole number arithmetic by subskill. To this end, the ASMD scorecard (p. 256) is scanned to note the cards with which a child is having difficulty, and Table 1 is referred to to note the subskills of whole number arithmetic which correspond to the cards. The rule of thumb for diagnosing is that two or more cards with a "miss" for a subskill indicate a weakness in the subskill.

For a ready source of prescriptions, see Shoecraft, P.J. BASIC MATHEMATICS: A BLUEPRINT FOR SUCCESS, Addison-Wesley Publishing Company, Inc., 1979. The whole number portion of the table of contents of same is nearly identical to the contents of Table 1!

Table 1. The Correspondence between the ASMD Cards and the Subskills of Whole Number Arithmetic

ASMD cards	Subskills of Whole Number Arithmetic by Operation			
	Addition	**Subtraction**	**Multiplication**	**Division**
1-4	Addition facts	Subtraction facts	Multiplication facts	Division facts
5-8	Three single-digit addends	Exchanging facts	Multiplication facts	Division facts
9-12	Addition without exchanging	Subtraction without exchanging	Single-digit multiplier without exchanging	Single-digit divisor, two-digit dividend
13-16	Two addends with exchanging of ones	Subtraction with exchanging of tens	Single-digit multiplier with exchanging of ones	Single-digit divisor, three-digit dividend
17-20	Two addends with exchanging of tens	Subtraction with exchanging of hundreds	Single-digit multiplier with exchanging of tens	Single-digit divisor, no zeros in quotient
21-24	Two addends with exchanging of ones and tens	Subtraction with exchanging of tens and hundreds	Single-digit multiplier with exchanging of ones and tens	Single-digit divisor, zeros in quotient
25-28	Three addends with exchanging of ones and tens	Subtraction with exchanging of tens, hundreds, thousands, ...	Single-digit multiplier with exchanging of ones, tens, hundreds, ...	Two-digit divisor, no zeros in quotient
29-32	Addition with exchanging of ones, tens, hundreds, ...	Exchanging across zeros	Two-digit multiplier	Two-digit divisor, zeros in quotient

ASMD Scorecard (p. 256)

A record keeping instrument for the problems on the ASMD cards.

The purpose of the ASMD scorecard is to assist with the diagnosis of weaknesses in whole number arithmetic as explained under **ASMD Cards**. To this end, have children use the same scorecard every time they play **Doll, Doll House, World Series,** or **Super Bowl,** and have them work each problem on the **ASMD cards** once only. The objective is to have them fill in the parts of the scorecard which pertain to the subskills of whole number arithmetic which they should know already or are learning about. A strategy for accomplishing this is to have them play the games for two or three class periods that they might fill in the bulk of what they are to fill in, and then to have them work from only the **ASMD cards** for an additional class period that they might fill in the rest of what they are to fill in.

Attribute Shapes (pp. 14-15)

A set of manipulatives for sorting and ordering on the basis of size (big or little), color (red, blue, or yellow), and shape (square, triangle, circle, or rectangle). A classroom set of them would consist of one set per child.

A major use of the **Attribute Shapes** is to develop the concept of number by having children sort them and order them: sort them into, say, 2, 3, or 4 piles on the basis of size, color, and shape, respectively, and order them by building "sameness trains" like the ones on page 17. (A sameness train is a string of "cars" for which the cars next to one another are the same in a predetermined number of ways.) The objective of the sorting is to provide a basis for cardination or "How many?" questions: How many big ones? Yellow ones? Triangles? That of the ordering to provide a basis for ordination or "Which one?" questions: Which is the first one? The second one? The third one?

Another major use of the **Attribute Shapes** is to give meaning to the union, intersection, and complement of sets by having children sort them as with the **Two-** and **Three-Way Sort task cards** on pages 23 and 24, respectively. The union of sets is the combination of whatever the sets consist of. The intersection of sets whatever the sets have in common. And the complement of sets whatever the sets do **not** consist of. Some examples of the union, intersection, and complement of sets are given in the instructions for the task cards.

Balance Beam Cutout (p. 129)

The beam for a balance beam. A classroom set of balance beams would

consist of one per child.

To make a balance beam with the beam, suspend the beam from a large paper cup as shown in Figures 3 and 4. (An alternative to a large paper cup is an empty milk carton.)

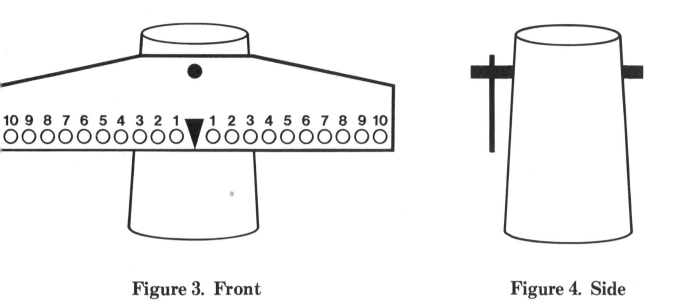

Figure 3. Front   Figure 4. Side

If the beam is slightly off balance to begin with, trim the heavy end of the beam with scissors. To weight the beam, use partially straightened paper clips as shown in Figure 5.

Figure 5

The major use of a balance beam is to develop the concept of equality by having children use one, one per child, to solve problems like the ones on pages 130-132. Since a balance beam is self-correcting, and the problems are non-verbal, even very young children can solve them and, in doing so, not only experience equality in a meaningful way, but are introduced to the basic facts for addition and multiplication.

Exploring equality on a balance beam

Another use of a balance beam is to motivate problem solving by having children use one, one per child, to solve teacher-directed problems like the following:

Addition:

"**Motley Crab Adder** found three and then found seven more. How many did he find altogether? To find out, put a three and a seven on one side of the beam and balance the beam with one weight."

Subtraction:

"Ork is nine. Zlu is six. Ork is how much older than Zlu? To find out, put a nine on one side of the beam and a six on the other side of the beam and balance the beam with one weight."

Multiplication:

"**Sir Crab Multiplier** found eight two times. How many did he find altogether? To find out, put two eights on one side of the beam and balance the beam with a 10 and one other weight."

Division:

"A truck can carry four. How many truck loads to carry 20? To find out, put 20 on one side of the beam and balance the beam with fours only."

If possible, precede any work with a balance beam with a play period on a teeter totter where small children are asked to balance large children and one child is asked to balance several children. In responding, they will discover the principle crucial to the working of a balance beam, that weight times distance equals weight times distance. They will also experience what equality "feels" like.

Exploring equality on a teeter totter

To appreciate the importance of having children work with a balance beam, ask some first or second graders what they think of the following equations: $5 = 2 + 3$, $7 = 7$, and $1 + 9 = 4 + 6$. They will probably reject them, insisting that $5 = 2 + 3$ should be rewritten as $2 + 3 = 5$, that $7 = 7$ should be rewritten as $7 + 0 = 7$, and that $1 + 9 = 4 + 6$ should be rewritten as $1 + 9 = 10$ and $4 + 6 = 10$. In other words, they will probably **not** view the equal sign as meaning

"the same as." Instead, they will view it as a sort of punctuation for th
location of "the answer." Having them work with a balance beam can hel
remedy this.

Balance Beam Worksheets (pp. 130-132)

Non-verbal directives for the balance beam. (See BALANCE BEAM CUT-OUT.)

The convention for the worksheets is as follows:

A triangle between weights represents the balance point of a balance beam.

A triangle without a slash means the balance beam is balanced.

A triangle with a slash means the balance beam is tilted.

Weights next to one another represent weights on different (but not necessarily adjacent) hooks.

Overlapping weights represent weights on the same hook.

Weights with numbers on them represent weights on hooks bearing the numbers.

On page 130, problem b can be used to introduce **even numbers**, problem c, **odd numbers**, and problem e, the **distributive property**, the fact that $a(b + c) = ab + ac$ for all numbers **a**, **b**, and **c**. On page 131, problem c can be used to introduce **less than** and **greater than**, but perhaps a better way to introduce them is as "greedy" mouths: mouths that always eat the bigger number. And on page 132, problems a and b can be presented in an exhaustive sense: How many solutions? The answer: Only the number indicated.

Regarding the problems for which there are only the indicated number of solutions, an assertion to that effect is easily verified by putting all the number of solutions indicated on a balance beam and noting the pattern they exhibit. They show that the "next" solution, if there were one, would extend beyond the limits of the balance beam.

The answers for the worksheets will vary, but they will be something like the following:

Page 130:

a. $2 + 1 = 3$
$5 + 3 = 8$
$9 + 1 = 10$

b. $2 \times 1 = 2$
$2 \times 4 = 8$
$2 \times 5 = 10$

c. $2 \times 2 + 1 = 5$
$2 \times 3 + 1 = 7$
$2 \times 4 + 1 = 9$

d. $3 \times 1 = 1 + 2$
$3 \times 4 = 5 + 7$
$3 \times 6 = 8 + 10$

e. $2 \times 3 = 2 \times 1 + 2 \times 2$
$2 \times 6 = 2 \times 2 + 2 \times 4$
$2 \times 10 = 2 \times 3 + 2 \times 7$

Page 131:

a. $2 + 1 = 1 + 2$
$8 + 7 = 5 + 10$
$6 + 3 = 4 + 5$

b. $3 + 2 + 1 = 6$
$4 + 3 + 2 = 9$
$6 + 3 + 1 = 10$

c. $1 < 2, \quad 3 > 2$
$3 < 6, \quad 5 > 1$
$4 < 10, \quad 9 > 4$

d. $3 + 2 + 1 = 2 \times 3$
$5 + 4 + 1 = 2 \times 5$
$9 + 8 + 3 = 2 \times 10$

e. $5 + 4 + 3 = 2 + 10$
$7 + 6 + 1 = 6 + 8$
$8 + 7 + 4 = 9 + 10$

f. $2 \times 3 = 3 \times 2$
$2 \times 6 = 3 \times 4$
$2 \times 9 = 3 \times 6$

Page 132:

a. $3 \times 4 = 4 \times 3$
$3 \times 8 = 4 \times 6$

b. $4 \times 5 = 5 \times 4$
$4 \times 10 = 5 \times 8$

c. $2 \times 3 = 2 + 4$
$2 \times 6 = 4 + 8$
$2 \times 9 = 8 + 10$

d. $3 \times 2 = 1 + 2 + 3$
$3 \times 5 = 2 + 3 + 10$
$3 \times 8 = 6 + 8 + 10$

e. $2 \times 2 = 4 \times 1$
$2 \times 4 = 4 \times 2$
$2 \times 6 = 4 \times 3$

Balancing Act

A counting game for two to four similar to Clown.

Materials:

One die and one Balancing Act gameboard (pp. 52-53) per group. One marker per player.

Directions:

To begin, the players put their markers on the Start arrow on the gameboard. Then each player rolls the die. The player rolling the highest number plays first. The play rotates clockwise.

To play, a player rolls the die and moves his or her marker along the trail on the gameboard as many spaces as the number showing on the die. If the player lands on an apple, the player counts the apples on the head of the character for that apple and moves as many more spaces as the number of apples. The play then goes to the next player.

The first player to reach the smiley face at the end of the trail wins.

Balloon Pop

A game for two on the subtraction facts. With different numbers, as with the blank Balloon Pop gameboard on page 156, a game on any of the basic facts.

Materials:

One die and one Balloon Pop gameboard (p. 155) per group. Ten markers per player.

Directions:

To begin, the players decide on who gets which of the two sets of 10 balloons each on the gameboard. Then each player rolls the die. The player rolling the higher number plays first.

To play, the players take turns rolling the die and "popping" each other's balloons by putting a marker on a balloon of the same number as the number showing on the die.

The first player to pop all of the other player's balloons wins.

Variation:

Have the players play with a consumable copy of the gameboard and color their own balloons instead of popping each other's balloons. Then make the winner the first player to color all of his or her balloons.

Bank It or Clear It

BANK IT

A game for two to four on exchanging "up" (regrouping) that assists with the development of the concepts of base and place value and gives meaning to addition of whole numbers.

Materials:

For each group, a die (or two dice) and a set of multi-base blocks or a four-color assortment of counters. For each player, a Bank It or Clear It gameboard (p. 99 for the blocks, p. 100 for the counters).

Directions:

In preparation, a land is decided on and, if using blocks, the units, longs, flats, and cubes for the land are garnered. If using counters, the gameboards are color coded, and the counter equivalents of the units, longs, flats, and cubes are garnered.

To begin, each player rolls the die. The player rolling the highest number plays first. The play rotates clockwise.

To play, a player rolls the die and places as many UNITS as the die indicates on his or her gameboard in the units column. Then the player "legalizes," that is, makes whatever exchanges are necessary to keep from breaking the law of the land and having to go to jail, and places the resultant on his or her gameboard in the appropriate columns. (As explained under The Great Legalizer, the law of the land in two land is "Never, never get caught with two or more things alike else GO TO JAIL!" The law of the land in three land is "Never, never get caught with three or more things alike else GO TO JAIL!" The law of the land in four land is "Never, never get caught with four or more things alike else GO TO JAIL!" And so on. Thus the law of the land in each land necessitates the exchanging of a given number of units for a long, the same number of longs for a flat, and the same number of flats for a cube.) The play then goes to the next player.

The first player to get a cube without going to jail wins.

To speed up the game in the "bigger" lands, and in ten land in particular, have players use two dice and play to only a flat.

Variation:

To emphasize base ten numeration with the game, put, say, 20 numbers between 1 and 100 on the blackboard. Then have the players play the game in ten land using two dice, and make the winner the first player to get a flat or make one of the numbers on his or her gameboard in the process.

This variation can be played with or without the gameboards and blocks or counters. If with, it makes numbers meaningful by relating them to the displays on the gameboards. If without, that is, with only the dice and paper and pen or pencil to keep track of what WOULD be on the gameboards if they and the blocks or counters WERE being used, it provides drill and practice on addition, but in a meaningful setting.

CLEAR IT

A game for two to four on exchanging "down" (regrouping) that assists with the development of the concepts of base and place value and gives meaning to subtraction of whole numbers.

Materials:

The same as for Bank It.

Directions:

In preparation, a land is decided on, and the materials for the land are garnered the same as for Bank It. Then each of the players puts the block or counter equivalent of one unit, one long, one flat, and one cube on his or her gameboard in the appropriate columns.

To begin, each player rolls the die. The player rolling the highest number plays first. The play rotates clockwise.

To play, a player rolls the die and REMOVES as many units as the die indicates from his or her gameboard. This will often require a player to exchange a cube for flats, a flat for longs, or a long for units. (Note: There is no going to jail here. As soon as a player gets too many of one thing alike, the player always gives enough of them away to keep from going to jail.) The play then goes to the next player.

The first player to clear his or her gameboard wins.

To speed up the game in the "bigger" lands, and in ten land in particular, have players use two dice and start with only one unit, one long, and one flat on their gameboards.

Variation:

To emphasize base ten numeration with the game, put, say, 20 numbers between 1 and 111 on the blackboard, and have the players play the game in ten land using two dice and starting with only one unit, one long, and one flat on their gameboards. Then make the winner the first player to clear his or her gameboard or make one of the numbers on his or her gameboard in the process.

As with the variation for Bank It, this variation can be played with or without the gameboards and blocks or counters. If with, it makes numbers meaningful by relating them to the displays on the gameboards. If without, it provides drill and practice on subtraction, but in a meaningful setting.

Blank Cards

Blank cards for adding to the game and activity cards in this book.

The blank cards on page 6 are for adding to the **Action cards**, **ASMD cards**, and **Shopping cards** and the cards for

✔ Addition Facts Rummy

✔ Doll

- ✔ Doll House

- ✔ Metric Concentration

- ✔ Multiplication Facts Rummy

- ✔ Pick-a-Pair

- ✔ Place Value Rummy

- ✔ Shape Rummy

- ✔ Super Bowl

- ✔ World Series

The blank cards on page 7 are for adding to the **Word cards** and the cards for **Ziggy's Home Run.** And the blank cards on page 8 are for adding to the **Digit cards, Division cards,** and the cards for **Gimmi** and **Telephone.**

No blank cards were included for the **Numeral cards** and **Sequence cards** because of the ease with which the cards for such can be made.

Blank Dominoes (p. 10)

Blank dominoes for adding to the dominoe games in this book:

- ✔ Addition and Subtraction Dominoes

- ✔ Double Nine Dominoes

- ✔ Fraction Dominoes

- ✔ Geo Dominoes

- ✔ Numeral Dominoes

Blank Tags (p. 9)

Blank tags for adding to the tags for Ollie Octopus.

Block It or Shade In

BLOCK IT

A game for two on the numeric properties of the colored rods.

Materials:

One die, a box of colored rods, and one Block It or Shade In gameboard per group. The small gameboard (p. 38) is for short games, the large gameboard (p. 39) for long games.

Directions:

To begin, each player rolls the die. The player rolling the higher number plays first.

To play, the players take turns rolling the die and covering the squares on the gameboard with a rod as long in units as the number showing on the die. The rod must lie entirely within the boundaries of the gameboard. If a player cannot fit a rod on the gameboard, the player loses that turn.

The play continues until the gameboard is completely covered. The player to make the last play wins.

To speed up the game with the larger gameboard, have players use two dice and allow combinations of rods (e.g., a 10-rod and a 2-rod for a 12).

SHADE IN

A game for two on making one-to-one correspondences. Also, a game on the concept of area.

Materials:

The same as for Block It except with two crayons of different colors instead of the box of colored rods.

Directions:

To begin, each player rolls the die. The player rolling the higher number plays first.

To play, the players take turns rolling the die and coloring as many squares on the gameboard that share a common side as the number showing on the die. The squares need not form a rectangle. If a player cannot color the number of squares indicated, the player loses that turn.

The play continues until all the squares are colored. The player to color the most squares wins.

To speed up the game with the larger gameboard, have players use two dice.

Boat

A game for two to four on the numeric properties of the colored rods and the addition facts up to sums of ten.

Materials:

Two dice and a box of colored rods per group. One Boat gameboard (p. 141) per player.

Directions:

Playing Boat

To begin, each player rolls the dice. The player rolling the highest total plays first. The play rotates clockwise.

To play, the players take turns rolling the dice and putting one or two colored rods on their gameboards in the spaces provided. If one, the rod must correspond to one of the numbers showing or to the sum of the numbers showing. If two, one of the rods must correspond to one of the numbers showing, and the other rod must correspond to the other number showing. To illustrate, if a player rolls a 3 and a 4, the player may put a 3-, 4-, or 7-rod or a 3-rod AND a 4-rod on his or her gameboard.

Only rods that fit the spaces on the gameboards exactly can be put on the gameboards. If a player cannot put a rod on his or her gameboard, the player loses that turn.

The first player to cover the boat on his or her gameboard wins.

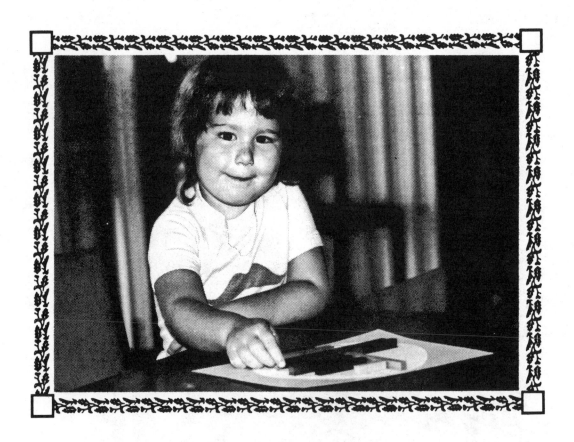

A winner every time

Bridge

A game for two similar to Lily Pad on making one-to-one correspondences and putting things next to one another.

Materials:

One Bridge gameboard (p. 46) and 10 bridge bits (p. 47) per player. Some envelopes for the bridge bits.

Directions:

In preparation, the bridge bits are packaged in varying amounts in the envelopes.

To play, the players take turns taking envelopes and putting the bridge bits therein on their gameboards in the spaces provided, one bridge bit per space. The bridge bits are to be placed next to one another starting at one side of the gorge on the gameboard and proceeding to the other side.

The first player to build a bridge across the gorge wins.

Playing Bridge

Bubbles (p. 74)

A numeration activity on the number names from 1 to 10, inclusive.

Build a Cube or Break a Cube

BUILD A CUBE

A variation of Bank It on exchanging "up" (regrouping) with an emphasis on numeration.

Materials:

For each group, a die (or two dice), an assortment of multi-base blocks (or counter equivalent of such), and a Build a Cube or Break a Cube gameboard (p. 101).

Directions:

The same as for Bank It except that the goal is always to be the first to "build a cube," that is, to exchange for a cube. Thus for the "bigger" land, such as five land, six land, or seven land, two dice are used, and the gameboard is prepared differently. Each player selects a side of the gameboard and, if playing the game in five land, starts the game with four flats in his or her 100 space. If in six land, with five flats in his or her 100 space. If in seven land, with six flats in his or her 100 space. And so on, with the number of flats always being one less than the number of flats needed for a cube. In this way the game can be played to completion within a reasonable amount of time.

Variation:

To emphasize base ten numeration with the game, put, say, 20 numbers between 900 and 999 on the blackboard. Then have the players play the game in ten land using two dice and starting with nine flats in each of their 100 spaces, and make the winner the first player to build a cube or make one of the numbers on his or her side of the gameboard in the process.

BREAK A CUBE

A variation of Clear It on exchanging "down" (regrouping) with an emphasis on numeration. (See Bank It or Clear It.)

Materials:

The same as for Build a Cube.

Directions:

The same as for Clear It except that the goal is to be the first to "break a cube." Thus to win, a player has only to have to exchange his or her cube for some flats rather than to have to completely clear his or her side of the gameboard.

Variation:

The same as for Build a Cube except with numbers between 1000 and 1111 on the blackboard.

Caboose

A game for two to four on the multiplication facts for which both factors are less than or equal to six.

Materials:

Two dice and a box of colored rods per group. One Caboose gameboard (p. 88) per player.

Directions:

To begin, each player rolls the dice. The player rolling the highest total plays first. The play rotates clockwise.

To play, the players take turns rolling the dice and putting colored rods on their gameboards in the spaces provided. The rule is that the rods must correspond to the product of the numbers showing on the dice in one of three ways: as the one number amount of the other number in rods, as the other number amount of the one number in rods, or as a single rod the same in length as the product of the numbers. To illustrate, if a player rolls a 2 and a 3, the player may put two 3-rods, three 2-rods, or a 6-rod on his or her gameboard. No other possibilities are allowed.

Only rods that fit the spaces on the gameboards exactly can be put on the gameboards. If a player cannot put all the rods possible on his or her gameboard, the player puts what he or she can on his or her gameboard. If a player cannot put any rods on his or her gameboard, the player loses that turn.

The first player to cover the caboose on his or her gameboard wins.

Car Park

A game for two similar to Watermelon on making one-to-one correspondences.

Materials:

One Car Park gameboard (p. 25) and 10 Car Park cars (p. 26) per player. Some envelopes for the cars.

Directions:

In preparation, the cars are packaged in varying amounts in the envelopes.

To play, the players take turns taking envelopes and "parking" the cars therein on their gameboards in the parking spaces provided, one car per parking space.

The first player to fill up the car park on his or her gameboard wins.

Variation:

The same game except with pegboard strips in place of the gameboards and pegs in place of the cars.

Circus (p. 206)

A worksheet on the division facts. With different numbers, as with the blank Circus on page 207, a worksheet on any of the basic facts.

Clown

A counting game for two to four similar to Balancing Act.

Materials:

One die and one Clown gameboard (pp. 54-55) per group. One marker per player.

Directions:

To begin, the players put their markers on the Start arrow on the gameboard. Then each player rolls the die. The player rolling the highest number plays first. The play rotates clockwise.

To play, a player rolls the die and moves his or her marker along the trail on the gameboard as many spaces as the number showing on the die. If the player lands on a balloon, the player counts the balloons held by the clown for that balloon and moves as many more spaces as the number of balloons. The play then goes to the next player.

The first player to reach the smiley face at the end of the trail wins.

Collect-a-Shape

A counting game for two to four similar to Safari.

Materials:

One die and one Collect-a-Shape gameboard (p. 59) per group. One marker and one set of Collect-a-Shape shapes (p. 60) per player.

Directions:

In preparation, all the shapes are combined and put in a tray of some sort to where they can be gotten to easily.

To begin, the players put their markers on the Start arrow on the gameboard. Then each player rolls the die. The player rolling the highest number plays first. The play rotates clockwise.

To play, a player rolls the die and moves his or her marker along the arrows and shapes on the gameboard in accordance with the number showing on the die. If the player lands on an arrow, the play goes to the next player. If on a shape, the player takes a shape like it, and the play goes to the next player.

The first player to get four different shapes wins.

Color the Shapes (p. 278)

A worksheet on identifying and coloring shapes.

The worksheet is straightforward except for the "cloud" in the upper left hand corner of the worksheet. The cloud raises the issue of whether or not colors can overlap.

The worksheet relates well to identifying shapes in newspapers, magazines, and the environment.

Computasnake (p. 164)

A worksheet on the addition and subtraction facts. With different numbers, as with the blank Computasnake on page 165, a worksheet on any of the basic facts.

Connect-a-Dot (pp. 87-90)

Four worksheets on ordination.

Countdown (p. 220)

A self-correcting worksheet on mixed facts.

The countdown for the "straight" rocket is 10, 9, 8, . . . , 1. For the "crazy" rocket, 10, 6, 4, 5, 9, 3, 8, 2, 7, 1.

Count the Shapes (p. 279)

A worksheet on shape recognition.

Careful! The worksheet exhibits more shapes than is readily apparent: at least 7 squares, 19 triangles, 4 rectangles, 14 trapezoids, and 2 parallelograms.

Crossnumber Puzzles

Self-correcting exercises on the multiplication and division facts.

The exercises on pages 183 and 184 are on multiplication, and the exercise on page 214 is on division. Exercises like them can be made using the grid paper on page 314.

The answers to the exercises are as follows:

Page 183:

(Top)

(Bottom)

Page 184:

Page 214:

Digit Cards (pp. 120-121)

Cards with numbers in the thousands on them.

The **Digit cards** are used with **Hiking** to focus attention on the one's place, ten's place, hundred's place, and thousand's place.

Dinosaur Eggs (p. 162)

A puzzle on the addition and subtraction facts. With different numbers, as with the blank Dinosaur Eggs on page 163, a puzzle on any of the basic facts.

The egg bits on **Dinosaur Eggs** are to be cut out and put together to form three eggs. The egg bits with the same answers go together.

Division Cards (pp. 210-211)

Introduction type problems on division of whole numbers.

The **Division cards** are used with **Ski Slope** to provide drill and practice on the division facts and division with remainders.

Division Worksheet (p. 231)

An instructional aid for introducing the division algorithm.

The worksheet is used to explain division in terms of separating "neatly" (into twos, threes, fours, and so on). To motivate the worksheet, use it in conjunction with **The Impeccable Twin Dividers**. Then relate what is done on the worksheet to what is done in the division algorithm.

The steps in using the worksheet are as follows:

1. Write the problem in the boxes at the top of the worksheet.

2. Circle the number of hundreds, tens, and ones in the dividend.

3. Separate the number of hundreds that are circled into groups the size of the diviser.

4. Write the number of groups of hundreds in the answer space above the hundreds.

5. Exchange each of any hundreds left over for 10 tens.

6. Repeat steps 3 through 5 for the tens, then for the ones.

7. Write any ones left over in the remainder space provided.

An example of how $4\overline{)935}$ would be worked on the worksheet is given below.

Doll

A game for two to four on whole number arithmetic.

The significance of Doll is twofold: First, it reviews nearly all of whole number arithmetic. Second, it diagnoses individual strengths and weaknesses in the subject by subskill. The manner in which it diagnoses is explained under ASMD Cards and ASMD Scorecard.

Materials:

For each group, a set of Doll cards (p. 240), a deck of ASMD cards (pp. 251-254), and an ASMD key (p. 255). For each player, a Doll gameboard (p. 236), a set of Doll clothes -- a set for each doll -- (pp. 237-239), an ASMD scorecard (p. 256), and a pen or pencil and some scrap paper.

Directions:

In preparation, the Doll clothes are laid out near the gameboards, and the Doll cards and ASMD cards are shuffled together and laid face down near the gameboards. Then the players decide on who is to play first and the order of play.

To play, a player draws a card. If it is a Doll card, the player does what it says. If it is an ASMD card, the player works one of the problems on the card and checks his or her answer to the problem using the ASMD key. If incorrect, the player records a "miss" in the appropriate space on his or her scorecard, and the play goes to the next player. If correct, the player records a "hit" instead, takes an item of clothing for one of the dolls on his or her gameboard, and the play goes to the next player.

The first player to dress all three dolls on his or her gameboard wins.

Doll House

A game for two to four similar to Doll on whole number arithmetic.

Materials:

For each group, a set of Doll House cards (p. 246), a deck of ASMD cards (pp. 251-254), and an ASMD key (p. 255). For each player, a Doll House gameboard (pp. 242-243), a set of Doll House furniture (pp. 244-245), an ASMD scorecard (p. 256), and a pen or pencil and some scrap paper.

Directions:

The same as for Doll except that furniture instead of clothing is put on the gameboards.

The first player to furnish all six rooms on his or her gameboard wins.

Dot Paper (pp. 264-265)

A page of 2-centimeter dot paper and a page of 1-centimeter dot paper.

Dot paper is typically used to develop the concept of area, the realization that when we say that the area of a figure is, say, 10, that what we mean is that to cover the figure would take 10 of some other figure. It can also be used to give meaning to the solution of arithmetic word problems.

To develop the concept of area with dot paper, have children connect the dots on the paper to make a polygon such as a square, rectangle, or triangle. Then have them find the area of the polygon using one of two methods, the "wholes and halves" method or the "rectangle" method, as explained below.

The wholes and halves method of finding area amounts to little more than viewing a figure in terms of units and half-units of area and counting the units and half-units as shown below.

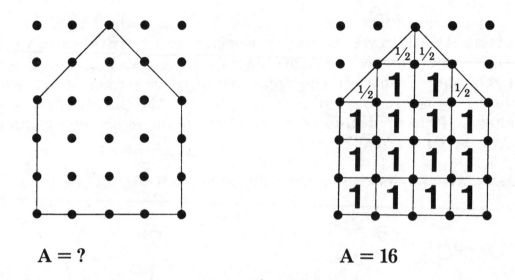

A = ?

A = 16

The value of the wholes and halves method of finding area is that it emphasizes the covering aspect of area and can be used by very young children. Its major drawback is that it works for only the simplest of figures made on dot paper, figures with horizontal, vertical, or main (45°) diagonal sides.

The rectangle method of finding area involves three steps as shown below.

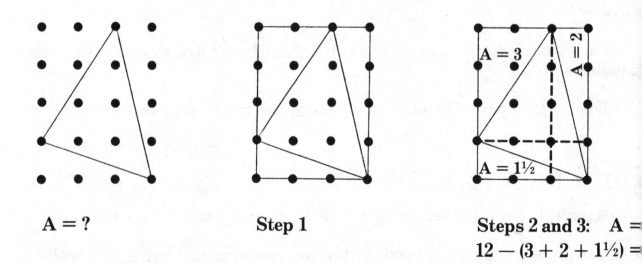

A = ?

Step 1

Steps 2 and 3:   A = 12 − (3 + 2 + 1½) = 5½

In Step 1, the triangle, the area of which was wanted, was enclosed in a "special" rectangle, a rectangle with horizontal and vertical sides which enclosed the triangle exactly. In Step 2, the areas of the new figures formed by the inclusion of the special rectangle were found. (Note that in each case the area was half the area of a rectangle as indicated by the dotted lines.) And in

Step 3, the sum of the areas of the new figures formed was subtracted from the area of the special rectangle.

The value of the rectangle method of finding area is that it encourages thinking and, like the wholes and halves method of finding area, emphasizes the covering aspect of area. Also, it works for any figure made on dot paper. Its major drawback is that it is a bit slow and cumbersome to use.

To give meaning to the solution of arithmetic word problems with dot paper, have children circle the dots on the paper in response to directives which underscore the "action" aspects of the four operations, the fact that in arithmetic word problems, addition corresponds to an implied combining action, subtraction to an implied separating action, multiplication to an implied combining "neatly" (by twos, threes, fours, and so on) action, and division to an implied separating "neatly" (into twos, threes, fours, and so on) action. An example of how this is done is given below.

Addition:

"Show me (Circle) two (as at the top in Figure 6). Show me (Circle) three (as at the bottom in Figure 6). Show me what **Motley Crab Adder** would do with them." (He would combine them as illustrated in Figure 7.)

Figure 6   Figure 7

"How would you describe what he would do?" ("He would JUST combine them.")

"How would you record what he would do?" (2 + 3 = 5)

Subtraction:

"Show me (Circle) eight (as in Figure 8). Show me what **The Scruffy Twin Subtractors** would do with it if they wanted five." (They would separate it as illustrated in Figure 9.)

Figure 8

Figure 9

"How would you describe what they would do?" ("They would JUST separate it.")

How would you record what they would do?" (8 − 5 = 3)

Multiplication:

"Show me (Circle) five fours (as in Figure 10). Show me what **Sir Crab Multiplier** would do with them." (He would combine them as illustrated in Figure 11.)

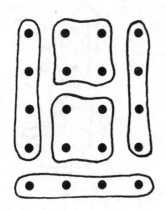

Figure 10

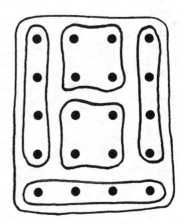

Figure 11

"How would you describe what he would do?" ("He would combine them 'neatly,' in this case by fours.")

"How would you record what he would do?" (5 × 4 = 20)

Division:

"Show me (Circle) 13 (as in Figure 12). Show me what **The Impeccable**

Twin Dividers would do with it if they wanted threes." (They would separate it as illustrated in Figure 13.)

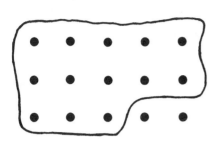

Figure 12

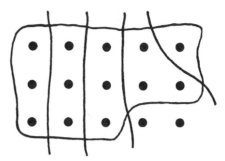

Figure 13

"How would you describe what they would do?" ("They would separate it neatly,' in this case into threes.")

"How would you record what they would do?" ($13 \div 3 = 4$ R1)

For best results, have them do the foregoing often, as often as once a week for a few minutes. Then teach them the "What's happening? How's it happening?" questioning technique explained under **Motley Crab Adder.** The technique is a strategy for extending an action-based understanding of the four operations to the solution of arithmetic word problems.

What is wanted are children who, if asked to close their eyes and think of the four operations, will see more than just the symbols $+$, $-$, $\times$, and $\div$ "written" across their foreheads, who will see, instead, a picture of what to look for in arithmetic word problems. The value in this objective is that its aim is to give children a way of thinking beyond the artificiality of arithmetic word problems, of thinking beyond the key words and catch phrases which abound in them. Its aim is to give them a way of thinking about "arithmetic word problems" as they will come to them when they are adults.

Double Nine Dominoes

In the "amateur" version, a game for two to four on making one-to-one correspondences. In the "professional" version, the same except on the multiplication facts.

Materials:

One set of Double Nine Dominoes (pp. 32-37) per group.

Directions (AMATEUR VERSION):

To begin, the dominoes are laid face down in what is called the bone yard and scrambled. Then the players take five dominoes each and examine them for doubles -- for dominoes with equivalent ends. The player with the highest double, or "spinner," lays it face up. This starts the game. The next player to play is the one to the left of the one playing the spinner. The play rotates clockwise. If no double is drawn, everything starts over until a double is drawn.

To play, a player matches the end of one of his or her dominoes with the end of a dominoe laying face up. If playing a double, the player matches the SIDE of the double with the end of the dominoe, that is, the player plays the double sideways. Conversely, if the dominoe laying face up is a double, the match is made with the side of the double. However, if the dominoe laying face up is the spinner, the match is made with the ends OR sides of the spinner. Thus there are never more than four possible plays: the plays off the ends or sides of the dominoes emanating from the ends and sides of the spinner. (See Figure 14.)

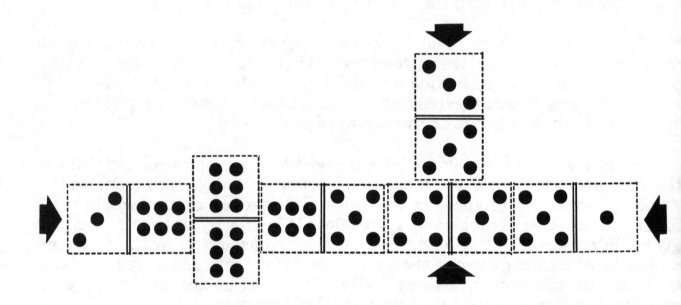

Figure 14. A double five for a spinner and, as indicated with arrows, four possible plays: a 1, a 3 (in two places), and a 5.

Once a match has been made, the play goes to the next player. If a player cannot make a match, the player draws from the bone yard until he or she can and does make a match.

The objective is to run out of dominoes. The first player to do so gets the points from the dominoes still with the other players -- a point for each dot.

If a player cannot make a match, and the bone yard is depleted, the play goes to the next player, and the object becomes to run out of dominoes or to have as few points on one's dominoes as possible. If no one runs out of dominoes, the player with the least number of points on his or her dominoes gets the points from the dominoes still with the other players minus the points from his or her dominoes.

The first player to get 100 points wins.

Variation:

To provide drill and practice on the number names from zero to nine, have players play with combined sets of Double Nine Dominoes and Numeral Dominoes.

Directions (PROFESSIONAL VERSION):

The same as for the amateur version except that players may score points during the game, as well as at the end of the game, by making multiples of, say, five with the ends of the dominoes they leave to play on. A player making such a multiple gets the points the multiple represents. (This is the reason for playing doubles sideways, to make for higher scores.)

Two examples of scoring for multiples of five are given in Figures 15 and 16.

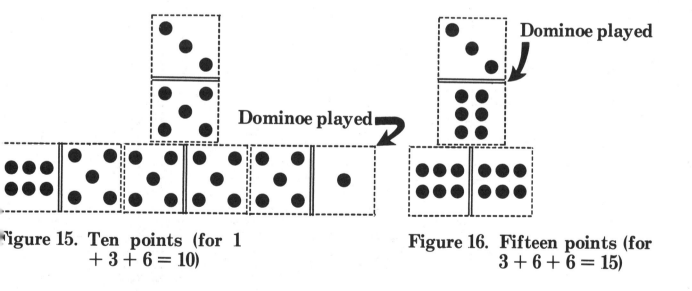

Figure 15. Ten points (for 1 + 3 + 6 = 10)

Figure 16. Fifteen points (for 3 + 6 + 6 = 15)

To exercise other multiplication facts, have players make multiples of numbers other than five.

Engine (p. 143)

A variation of Boat.

Experience Roster (p. 1)

An instrument for keeping track of the mathematics children experience.

An "experience roster" is of major value in an activity-oriented curriculum that allows for some option on what children work on. It lends order to a curriculum based on the familiar:

I hear and I forget.

I see and I remember.

I do and I understand.

An example of its use is given below. One or more check marks in a given column and row mean the experience in that column for the child in that row the number of times indicated.

Name	Experience — Sameness trains for Attribute Shapes	Sameness trains for Word cards	Car Park	Watermelon	Bridge	Lilly Pad												
B.J.				✓✓	✓✓												
Jess				✓✓	✓✓✓✓												
Lynne	✓	✓✓✓				✓											
Paul	✓✓		✓			✓											

Find the Cheese

A game for two on the concept of length.

Materials:

One die and one Find the Cheese gameboard (p. 300) per group. One pen or pencil per player.

Directions:

To begin, each player rolls the die. The player rolling the higher number plays first.

To play, the players take turns rolling the die and drawing a path on the gameboard as many units long as the number showing on the die. (A unit is the distance between two vertically or two horizontally adjacent dots.) The path may zig zag, but only in a vertical and horizontal fashion. It cannot veer diagonally. Also, it cannot retrace a portion of another path. It can, however, cross another path. If a player cannot draw a path, the player loses that turn.

The object is to get to the cheese on the gameboard with a series of connected paths. The player starting on the left side of the gameboard aims for the dot on the left side of the cheese, and the player starting on the right side of the gameboard aims for the dot on the right side of the cheese. To get to the cheese, the players must get to the dots for the cheese exactly.

The play continues until both players get to the cheese. The player to get to the cheese with the shorter path wins.

Fish (p. 110)

A numeration exercise on two-digit numeration, namely, that of 54 fish.

Fly, Fly Away (p. 227)

A problem solving activity on subtraction of whole numbers. For different numbers, use the blank Fly, Fly Away on page 228.

To lighten the balloon on **Fly, Fly Away** to exactly 1250, drop the sandbag weighing 297.

Fraction Dominoes

A game for two to four on the meaning of fractions.

Materials:

One set of Fraction Dominoes (pp. 272-274) per group.

Directions:

The same as for the amateur version of Double Nine Dominoes except with different scoring: five points for each dominoe left in a player's hand.

Some examples of some matches for the Fraction Dominoes would be the "1 / 2" and the half-shaded triangle, the "3 / 4" and "three-fourths," and the "1" and the completely shaded circle.

Fraction Rulers

An instructional aid for giving meaning to fractions.

Materials:

One transparency and as many copies as children of either version of Fraction Rulers (pp. 269-270). One overhead projector.

Procedure:

To begin, project the transparency and pass out the copies. Then alternately point to fractions on the transparency for the children to call out and call out fractions for the children to point to on the copies.

Frog (p. 133)

A worksheet on the addition facts. With different numbers, as with the blank Frog on page 134, a worksheet on any of the basic facts.

Geo Dominoes

A game for two to four on matching shapes.

Materials:

One set of Geo Dominoes (pp. 280-282) per group.

Directions:

The same as for the amateur version of Double Nine Dominoes except with different scoring: five points for each dominoe left in a player's hand.

Some examples of some matches for the Geo Dominoes would be any circle and any other circle, any triangle and any other triangle, and any square and the word SQUARE.

Gimmi

A game for two to four on combining attributes.

Materials:

For each group, a set of the Attribute Shapes on pages 14 and 15 or the Word cards on pages 18-20 and a set of Gimmi cards for whichever: page 16 for the Attribute Shapes, page 21 for the Word cards.

Directions:

In preparation, the Attribute Shapes or Word cards are combined and put in a tray of some sort to where they can be gotten to easily. Then the Gimmi cards for whichever are laid face down in what is called the Gimmi pile and scrambled.

To play, the players take turns taking a Gimmi card from the Gimmi pile until they have enough cards to exactly describe one of the Attribute Shapes or Word cards. For either, this will require three cards. They then take whichever they can describe and return the cards by which the description was made to the Gimmi pile.

The first player to get five Attribute Shapes or Word cards wins.

Going to the Park

A game for two on classifying the People Pictures in relation to sex (male or female) and age (child or adult).

Materials:

One Going to the Park gameboard (p. 12) and one set of People Pictures (p. 11) per player.

Directions:

In preparation, each player lays his or her People Pictures face down and scrambles them.

To play, the players take turns picking a People Picture and taking it to one of the parks on their gameboard in keeping with the classification scheme

on the gameboard: All women go to the park with the paddle pool. All men t
the park with the swing set. All girls to the park with the sandbox. And al
boys to the park with the monkey bars.

The first player to get four People Pictures in a park wins.

For a different classification, use the blank Going to the Park gameboar
on page 13.

Playing Going to the Park

Grid Paper (pp. 313-314)

A page of 2-centimeter grid paper and a page of 1-centimeter grid paper.

A major use of grid paper is to find the areas of non-polygonal shapes such
as circles, hands, and feet. The shapes are drawn or traced on the grid paper
and the areas of the shapes are found by counting the squares and parts of
squares they encompass.

Hex

A game for two on putting things next to one another -- a skill prerequisite
to playing board games.

Materials:

One Hex gameboard (p. 50) per group. About 15 markers per player.

Directions:

To begin, the players decide on who goes first.

To play, the players take turns putting a marker on the gameboard on an unoccupied hexagon which is next to an occupied hexagon. If it is the very first play, the player puts it on, say, one of the "A" hexagons, after which the other player puts a marker on one of the "B" hexagons.

The first player to get from an A hexagon to the other A hexagon or from a B hexagon to the other B hexagon wins.

The game allows for more strategy than is readily apparent.

Hiking

A game for two to four on base ten numeration.

Materials:

One Hiking gameboard (pp. 118-119) and one set of Digit cards (pp. 120-121) per group. One marker per player.

Directions:

To begin, the players put their markers on the Start space on the gameboard. Then each player draws a Digit card. The player drawing the card with the highest number plays first. The play rotates clockwise.

To play, a player draws a Digit card and, in accordance with the place specified on the space the player is on, moves the number of spaces given on the card for that place. To illustrate, if a player draws the card bearing 4051 and is on the Start space specifying the thousand's place, the player moves four spaces. The play then goes to the next player.

The first player to reach the Home space on the gameboard wins.

Horse (p. 142)

A variation of Boat.

House (p. 186)

A variation of Caboose.

Hundred Chart (p. 178)

A list of the first 100 counting numbers. Also, an instructional aid with which to introduce multiplication, broach the subject of prime numbers, and provide drill and practice on the basic facts.

To use the **Hundred Chart** to introduce multiplication, make about 10 copies of the chart per child. Then, on one of the copies, have children color every second number, that is, all the multiples of 2, and look for patterns. Then have them repeat the process on the other copies for the multiples of 3, 4, 5, and so on up through 10. Finally, have them return to the copies already colored and color a different multiple of 2 through 10 on each one and look for patterns in the intersection of the multiples.

To use the **Hundred Chart** to broach the subject of prime numbers, have children circle the 2 -- the first prime number -- and cross out all the multiples of 2 from there on. Then have them circle the next number -- the 3 -- and cross out all the multiples of 3 from there on. Then have them circle the next number -- the 5 -- and cross out all the multiples of 5 from there on. And so on until the chart is exhausted. The numbers circled will be the prime numbers less than 100.

And to use the **Hundred Chart** to provide drill and practice on the basic facts, have children play **Hundo** as explained below:

HUNDO

A game for two to four on the basic facts.

Materials:

One deck of regular playing cards per group. One Hundred Chart per player.

Directions:

To begin, the players take the 2 through 10 cards from the deck of playing cards, shuffle the cards, and take five cards each. The players then decide on an operation: addition, subtraction, multiplication, or division.

To play, the players use the operation and the numbers on the cards to make as many of the numbers on the chart as they can. To illustrate, for the operation of multiplication and the numbers 2, 5, 5, 6, and 8, a player could make the 10 (2×5), 12 (2×6), 16 (2×8), 25 (5×5), 30 (5×6), 40 (5×8), 48 (6×8), 50 ($2 \times 5 \times 5$), 60 ($2 \times 5 \times 6$), 80 ($2 \times 5 \times 8$), and 96 ($2 \times 6 \times 8$). The players then total their numbers, and the player with the highest total wins.

To alter the complexity of the game, have players decide on two operations.

Improve Your Aim (p. 225)

A problem solving activity on addition of whole numbers. For different numbers, use the blank Improve Your Aim on page 226.

The least number of shots is three: $32 + 34 + 34 = 100$. A greater number of shots would be six: $11 + 11 + 19 + 19 + 20 + 20 = 100$.

Jellybeans (p. 109)

A numeration exercise on two-digit numeration, namely, that of 27 jellybeans.

Kilometer Count (p. 138)

A problem solving activity on simple sums.

The shortest path from the castle to the church is 13 kilometers. A variation on **Kilometer Count** is to have children make some maps of their own to give to one another to work out.

Lady Bugs (p. 159)

A related facts worksheet on the addition and subtraction facts. With different numbers, as with the blank Lady Bugs on page 160, a worksheet on any of the basic facts.

Lattice Multiplication (p. 230)

A worksheet for lattice multiplication.

Two examples of lattice multiplication are given below.

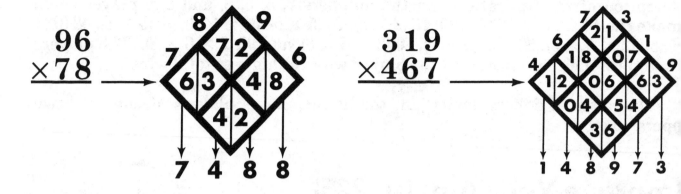

The procedure in each case was to write the problem across the top of the lattice, fill in the cells of the lattice **two at a time** with the products of the digits of the numbers across the top, and to add the columns of numbers in the cells. If you study the examples carefully, you will see how this was done.

Lattice multiplication is a method of multiplying which, with the pocket calculator encroaching on the conventional method of multiplying, is not necessarily a bad method to teach to children. Its most striking feature is that it keeps the multiplying separate from the adding. Also, it is artistic, keeps work neat, is easily learned, is not mentally tiring, and allows for diagnosis of errors. Its major drawback is that unless one already has a lattice, as with the worksheet for lattice multiplication, it takes time to draw the lattice.

As a follow-up to lattice multiplication, in place of teaching the conventional method of multiplying, consider teaching the "multiplication facts algorithm," a relatively new procedure for multiplying which has a number of advantages over the conventional method of multiplying. Two examples of the multiplication facts algorithm in use are given below. Note the sameness between them and the examples used for lattice multiplication: the same problems, the same setting out of numbers.

$$
\begin{array}{r}
96 \\
\times\,78 \\
\hline
28 \\
74 \\
32 \\
64 \\
\hline
7488
\end{array}
\qquad
\begin{array}{r}
319 \\
\times\,467 \\
\hline
173 \\
206 \\
864 \\
105 \\
246 \\
103 \\
\hline
148973
\end{array}
$$

The procedure in each example was to write the products **diagonally** and to underline the first digit in the first product for a digit in the multiplier. The reason for the underlining was to direct the placement of the first product for the next digit in the multiplier. Again, if you study the examples carefully, you will see how this was done.

The advantages of the multiplication facts algorithm over the conventional method of multiplying are much the same as the positive attributes of lattice multiplication: It separates the multiplying from the adding, is easily learned, partly because of the "one-two, one-two, . . ." rhythm that goes with writing the products, is non-stressful in that it allows a user to take a break anywhere in a problem and later on pick up where he or she left off, and is diagnostic in that it allows a teacher to see where in a problem things went wrong in the event of a wrong answer. But unlike lattice multiplication, it requires no more pen or pencil work than the conventional method of multiplying. Its only drawback, short of comparing it to a pocket calculator, is that it looks strange to anyone accustomed to the conventional method of multiplying.

Leaf Patterns (p. 176)

A skip-counting activity on the multiples of 2, 5, and 10. For different multiples, use the blank Leaf Patterns on page 177.

Lily Pad

A game for two similar to Bridge on making one-to-one correspondences and putting things next to one another.

Materials:

One Lily Pad gameboard (p. 48) and 10 "lily pads" (p. 49) per player. Some envelopes or match boxes for the lily pads.

Directions:

In preparation, the lily pads are packaged in varying amounts in the envelopes or match boxes.

To play, the players take turns taking envelopes or match boxes and putting the lily pads therein on their gameboards in the spaces provided, one lily pad per space. The lily pads are to be placed in sequence starting from the space nearest the frog and proceeding to the space nearest the bank.

The first player to fill in all the spaces on his or her gameboard wins.

Line Designs (pp. 296-299)

Four patterns for making line designs.

A **line design** is a design made by connecting points made to correspond to one another with lines to where the lines appear to be curved. The basis for such a design is a right angle whose sides meet two conditions: One, they are the same in length. Two, they exhibit the same number of equidistant dots. An example of such an angle and how it is made into a line design is given in Figure 17.

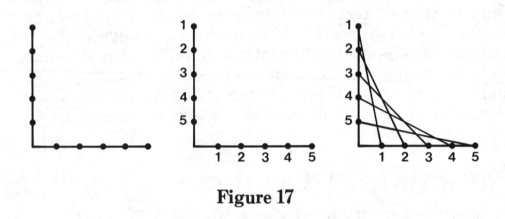

Figure 17

To alter the basic design, one varies on the relative lengths of the sides, the number of dots, and the size of the angle as illustrated in Figure 18.

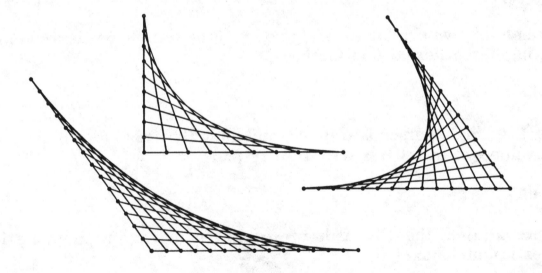

Figure 18

And to elaborate on the basic design, one combines angles as in Figure 19.

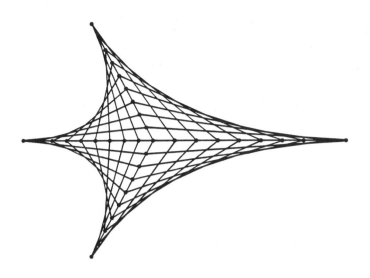

Figure 19

Note that an endless variety of line designs is possible.

The line designs for the four patterns for making line designs are as follows:

Page 296: Eye

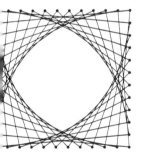

Page 297: God's Eye

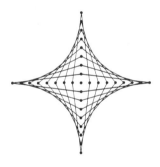

Page 298: Frisbee

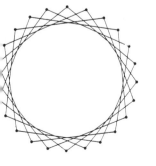

Page 299: Basket Ball

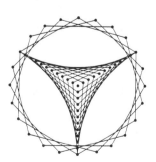

Liter Cutout (pp. 325-326)

A model of the liter. Also, a model of the ten-land cube.

The model focuses attention on the decimal nature of the metric system with the cutting and pasting involved in making it.

Man (p. 185)

A variation of Caboose.

Me as a Measure (p. 301)

A measuring activity in terms of non-standard units, namely, "body" units.

Me as a Measure and measuring activities like it are almost always delighted in by children. Moreover, they lead to the following worthwhile considerations: the advantages and disadvantages of measuring with one's body and the need for standard units of measure in modern society. The advantages would include accessibility in that we always have our bodies with us, the disadvantages the fact that we would occasionally be "shortchanged" in the market place depending on the size of the person doing the measuring.

Merry Measuring (p. 271)

A measurement activity on fractions.

Merry Measuring is used to motivate and give meaning to fractions. An interesting application deriving from it has to do with buying socks. If someone forgets their sock size, all they have to do is wrap a sock around their fist. One's foot length and fist circumference tend to be the same.

Materials:

One piece of string per child as long as each child.

Metric Concentration

A game for two to four on metric prefixes.

Materials:

One set of Metric Concentration cards (pp. 311-312) per group.

Directions:

To begin, the players decide on a "dealer." The dealer shuffles the cards and spreads them out face down to make a four-by-four array. The player to the left of the dealer plays first. The play rotates clockwise.

To play, a player draws two cards from the array. If the cards exhibit equivalent measurements, the player keeps the cards and continues to draw cards two at a time until he or she draws two cards which do not exhibit equivalent measurements, in which case he or she returns the cards which do not exhibit equivalent measurements to the array exactly as they were, and the play goes to the next player.

The key to determining equivalent measurements is knowing that kilo-means 1000, centi- 0.01, and milli- 0.001. Thus, relative to the cards,

1 m (meter) = 100 cm (centimeters)

The length of the line is 10 cm (centimeters).

1 cm (centimeter) = 10 mm (millimeters)

The volume of the cube is 1 mL (milliliter).

1 km (kilometer) = 1000 m (meters)

1 L (liter) = 1000 mL (milliliters)

1 kg (kilogram) = 1000 g (grams)

1 g (gram) = 1000 mg (milligrams)

The player with the most cards after all the cards have been paired wins.

Milliliter, 10-Milliliter, and 100-Milliliter Cutouts (p. 324)

Models of 1 milliliter, 10 milliliters, and 100 milliliters. Also, models of the ten-land unit, long, and flat.

The models focus attention on the decimal nature of the metric system with the cutting and pasting involved in making them.

Mirror Symmetry (pp. 294-295)

Two self-correcting worksheets on mirror symmetry.

A **mirror symmetric figure** is a figure with a "line of symmetry," a line along which a figure can be folded so as to make the figure coincide with itself or along which a mirror can be placed so as to reproduce in the mirror the half of the figure blocked out by the mirror. Two examples of mirror symmetric figures are given below. As indicated with dotted lines, the figure on the left has one line of symmetry, the figure on the right four lines of symmetry.

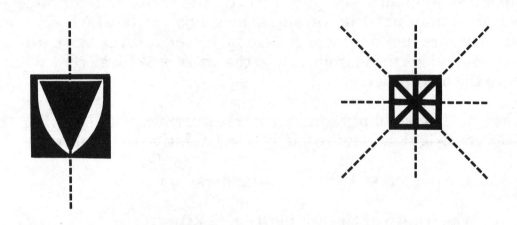

Exploring symmetry with a hand mirror

The answers for the worksheets are as follows:

Page 294:

Figures b, d, f, h, i, and j are mirrror symmetric. (Figure c would be mirror ymmetric, too, if it were not for the shading.)

Page 295:

As a follow-up to the worksheets, consider having children examine things such as plants, insects, and animals for mirror symmetry.

Monster (p. 137)

A self-correcting worksheet on the addition facts.

Moon Walk

A game for two on the addition, subtraction, multiplication, and division acts.

Materials:

For each group, a Moon Walk gameboard (p. 223) and three dice. For each player, four markers of a particular color.

Directions:

To begin, the players put their markers on Earth on the gameboard. Then each player rolls the dice. The player rolling the higher total plays first.

To play, the players take turns rolling the dice and trying to move a marker to one of the three destinations on the gameboard -- Starship, Lunar Station, or Moon -- by making one of the numbers on the destination with the numbers showing on the dice. To this end, the numbers on the dice may be added, subtracted, multiplied, or divided in any way possible so long as each number is used once and only once. To illustrate, if a player rolled a 1, 3, and 5 and wanted to make a number on Starship, the player could make the 1 $(5 - 3 - 1)$, 10 $((3 - 1) \times 5)$, or 12 $(3 \times (5 - 1))$.

Once a player makes a number on a destination, the player covers the number with a marker. And once a number is covered, it cannot be covered again until the marker on it is moved to the next destination. The rule is that a marker cannot be moved from one destination to the next until it is joined by its "fellow-traveler" markers. Thus a player cannot move a marker to Lunar Station until all of his or her markers are on Starship. And a player cannot move a marker to Moon until all of his or her markers are on Lunar Station.

The first player to get all four of his or her markers to Moon wins.

Motley and Mates (p. 263)

Miniatures of Motley Crab Adder, The Scruffy Twin Subtractors, Sir Crab Multiplier, The Impeccable Twin Dividers, The Great Legalizer, and The Magnificent Equalizer.

The major use of Motley and Mates is in making transparencies and task cards for arithmetic word problems.

Motley Crab Adder (pp. 257)

The personification of addition as combining.

Motley Crab Adder is used to help children understand arithmetic word problems as explained below and illustrated for Dot Paper and Shopping Cards. If colored, he makes an attractive item for the bulletin board. If enlarged, as with an overhead projector, and colored, he makes an attractive oster. And since he is independent of computational skill, he can be talked bout to even the youngest of children to help them see the mathematics mplicit in the world around them.

An introduction to Motley would go something like this:

"Meet **Motley Crab Adder,** a strange character indeed. He always combines things. That's just the way he is. No one knows why.

"People have talked to him about this and have told him that he is strange because of it. They have even suggested that he see a psychiatrist. But he just smiles and goes on combining things.

"Also, when he combines things, he combines them in just any old way. This shows in his sloppy appearance and is why his first name is Motley.

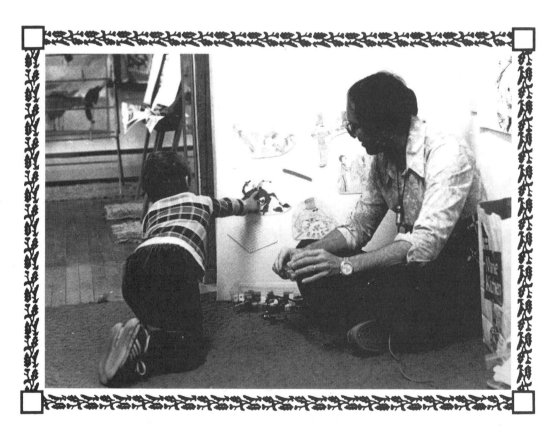

Who JUST combines things?

"And look at all the things he can do!" (Mime some of the following:

Raking grass or leaves

Saving money in a piggy bank

Putting letters in a mail box

Putting on make-up

Making a sandwich

"Can you show me these things he can do?" (Have children mime some of the following:

Putting a puzzle together

Putting toys in a toy box

Building a rock wall

Assembling something like a kite
or a train set

Packing a suitcase

Combining a 1 and a 2 to get a 3

To see how **Motley Crab Adder** helps children solve arithmetic word problems, consider him in relation to **The Scruffy Twin Subtractors** (p. 258), **Sir Crab Multiplier** (p. 259), and **The Impeccable Twin Dividers** (p. 260). Motley, the Scruffy Twins, Sir Crab, and the Impeccable Twins are personifications of the actions of combining, separating, combining "neatly" (by twos, threes, fours, and so on), and separating "neatly" (into twos, threes, fours, and so on), respectively -- of the kinds of actions taken with real things in the real world that can be interpreted mathematically: combining corresponds to addition, separating to subtraction, combining neatly to multiplication, and separating neatly to division.

The importance of Motley and mates is then seen in relation to the following four-step problem solving process for arithmetic word problems:

1. Understand the problem.

2. Decide what to do -- to add, subtract, multiply, or divide.

3. Do it.

4. Check the reasonableness of results.

Motley and mates help get things started. They help with Steps 1 and 2 in the problem solving process that whatever Step 3, paper-and-pencil or pocket calculator skills are known might be put to use.

To illustrate, suppose children were taught about addition as combining, subtraction as separating, multiplication as combining neatly, and division as separating neatly. Then to solve an arithmetic word problem, all they would have to do to get started is ask two questions:

1. What's happening, combining or separating?

2. How's it happening, just happening or happening neatly?

In so doing, they will come to see, for example, that throwing groceries into a shopping basket is combining, that making change is separating, that building a brick wall is combining neatly, and that sharing equitably with friends is separating neatly. In other words, that in related problems they should add, subtract, multiply, and divide, respectively.

The thing to keep in mind when helping children solve arithmetic word problems is that every combining action involves a separating action, and vice versa, just as every addition problem involves a subtraction problem, and vice versa. To illustrate, the action of combining groceries in a shopping basket presupposes the action of separating groceries from a shelf. Thus the key to helping children solve arithmetic word problems is to have them focus on the pertinent action.

To better appreciate Motley, the Scruffy Twins, Sir Crab, and the Impeccable Twins, they, in combination with **The Great Legalizer** (p. 261) and **The Magnificent Equalizer** (p. 262), illustrate the sum total of the ideas that are fundamental to the processes that are used over and over again in basic mathematics. (**The Great Legalizer** is the personification of the need in computing with whole numbers to regroup. **The Magnificent Equalizer** that of the need in working with fractions to make things "different yet the same" -- different in appearance, yet the same in quantity.) Thus upon introducing them to some kindergarten children, it would be mathematically correct to tell the children that for the next nine years in math they will review!

Multi-base Blocks

Patterns for making pictorial representations of multi-base blocks.

The large patterns (pp. 93-95) are for making paper models or flannel board replicas of multi-base blocks, the small ones (pp. 96-98) for making transparencies or task cards for multi-base blocks. For base ten, the patterns are cut out as is. For base two, base three, base four, and so on up to base nine only the relevant parts of the patterns are cut out. A classroom set of large

Exploring base and place value with multi-base blocks

blocks for base ten would consist of 10 units, 10 longs, 10 flats, and 5 cubes per child with the understanding that some of the material would have to be shared for some problems. A classroom set of large blocks for a base less than ten would consist of proportionately fewer units, longs, flats, and cubes.

The major use of multi-base blocks, in either paper or wooden form, is to develop the concepts of base and place value and give meaning to whole number arithmetic. To these ends, have children use them with **Bank It or Clear It, Build a Cube or Break a Cube, Addition in Different Lands,** and **Subtraction in Different Lands.**

Multiplication Facts Rummy

A game for two to four on the multiplication facts.

Materials:

One deck of Multiplication Facts Rummy cards (pp. 199-205) per group.

Directions:

The same as for Addition Facts Rummy except without runs.

An example of a spread for the Multiplication Facts Rummy cards would be the "2 × 3," "3 × 2," and "6." An example of a book the two-by-three array of hexagons as well.

Multiplication Facts Table (p. 175)

A list of all 100 multiplication facts.

The multiplication facts are unquestionably a tough bunch of facts to remember. They mark the mathematical demise of a lot of children. To help children learn the "hard" ones, the ones for which both factors are greater than or equal to six, show them the following finger method of multiplication as illustrated for 7 × 9:

> "Count to 7 on the fingers of one hand, putting down the fingers touched twice (in this case, 2 fingers). Then count to 9 on the fingers of the other hand, again putting down the fingers touched twice (in this case, 4 fingers). Now regard the fingers down -- all 6 of them -- as 10 each for a total of 60, multiply the 3 fingers up on the one hand by the 1 finger up on the other hand to get 3, and add the 60 and the 3 to get 63! To help you remember to multiply the numbers up, always click the

heels of your hands together before multiplying. It's not clear as to why this helps, but it does."

This method of multiplication was known by the Chinese more than 2000 years ago and works for all the multiplication facts from 6×6 to 9×9, inclusive. It works because of the following identity: $(5 + x)(5 + y) = 10(x + y) + (5 - x)(5 - y)$.

Non-digital Clocks

Pictures of non-digital clocks.

The large non-digital clock (p. 275) is for making paper models of same for demonstrations and class exercises on telling time. (e.g., "If this is 2:00 on my clock, show me 3:00 on your clocks.") And the small non-digital clocks (p. 276) are for individual work on telling time. For different times on the small non-digital clocks, use the blank **Non-digital Clocks** on page 277.

The non-digital clocks are included with the material on fractions that phrases such as "a quarter to" and "half past" might be given meaning. (e.g., A quarter to means a quarter or one-fourth of a clock face before the hour, and half past means one-half of a clock face past the hour.)

Number Houses (p. 161)

A related facts worksheet on addition and subtraction.

Numeral Cards (pp. 75-78)

Cards bearing the numerals 1, 2, 3, . . . , 12.

The **Numeral cards** are used to teach the numerals 1, 2, 3, . . . , 12. The procedures for using them are many and varied as illustrated below:

Have children match the cards from two sets of **Numeral cards.**

Have children put the **Numeral cards** in order.

Have children figure out the missing card or cards from a set of **Numeral cards.**

Have children throw two dice and select the **Numeral card** that corresponds to the total on the dice.

For each **Numeral card,** have children make a match with counters.

For each **Numeral card,** have children draw as many bees, birds, flowers, or the like as the number on the card.

Have children take turns holding up a **Numeral card** for one another to clap out.

Have children take turns hiding a **Numeral card** for one another to guess.

Have children play **Numeral Bingo** as explained below:

NUMERAL BINGO

A game for two to four on the numerals 1, 2, 3, . . . , 12.

Materials:

One set of Numeral cards per player.

Directions:

In preparation, the players elect a group leader. The group leader then shuffles his or her cards as each of the other players chooses six cards from his or her set of Numeral cards and spreads them out face up.

To play, the group leader turns over one of his or her cards and asks if anyone has a number like it showing. If so, they turn it over.

The first player to turn over all six of his or her cards wins.

Numeral Dominoes

In the "amateur" version, a game for two to four on the numerals 0, 1, 2, . . . , 9. In the "professional" version, the same except on the multiplication facts.

Materials:

One set of Numeral Dominoes (pp. 81-86) per group.

Directions:

The same as for either version of Double Nine Dominoes.

Some examples of some matches for the Numeral Dominoes would be any "1" and any other "1," any "5" and any other "5," and any "9" and the word NINE. Note: The Numeral Dominoes are compatible with the Double Nine Dominoes and the Addition and Subtraction Facts Dominoes.

Numeral Puzzles (pp. 67-69)

Self-correcting puzzles on the numerals 1, 2, 3, . . . , 12. For different numerals, or for the words instead of the symbols for the numerals, use the blank Numeral Puzzles on pages 70-72.

Putting the Numeral Puzzles together

Numeration in Different Lands (p. 102)

An exercise on the number name for a quantity depending on the land (base) it is in.

The numeration worksheet is used as a follow-up to **Bank It** and **Build a Cube** to connect the concepts of base and place value to numeration. It is easily completed with the help of **The Great Legalizer** as a reminder to not go to jail (to not fail to regroup).

The entries for the table on the worksheet are as follows:

Number of units	Two land	Three land	Four land	Five land	Six land	Seven land	Eight land	Nine land	Ten land
1	1	1	1	1	1	1	1	1	1
2	10	2	2	2	2	2	2	2	2
3	11	10	3	3	3	3	3	3	3
4	100	11	10	4	4	4	4	4	4
5	101	12	11	10	5	5	5	5	5
6	110	20	12	11	10	6	6	6	6
7	111	21	13	12	11	10	7	7	7
8	1000	22	20	13	12	11	10	8	8
9	1001	100	21	14	13	12	11	10	9
10	1010	101	22	20	14	13	12	11	10

Ollie Octopus

A game for two to four on base ten numeration. With different numbers, s with the blank tags on page 9 and the blank Ollie Octopus card on page 126, game on any of the basic facts.

Materials:

One set of Ollie Octopus tags (p. 127) per group. One Ollie Octopus card p. 122-125) per player.

Directions:

To begin, each player draws a tag. The player drawing the tag with the ighest number is the "group leader." All the tags are then laid face down and crambled.

To play, the group leader turns up a tag as each player, including the group eader, tries to match the tag with a leg of the octopus on his or her card. The rst player to do so gets to cover the leg with the tag. The group leader then urns up another tag, and the play continues as before. If no match can be made, the group leader keeps turning up tags until a match can be made.

The first player to cover all eight legs of the octopus on his or her card ins.

One Less (p. 174)

An integration of math and language arts on "knocking down" (making one less).

Knocking down is a skill taught to children in teaching them the addition facts. First, they are taught the "doubles" $(1 + 1 = 2, 2 + 2 = 4, 3 + 3 = 6, \ldots)$. Then they are shown how knowing how to "knock down" gives them a great many more addition facts. (For example, "If $2 + 2 = 4$, what's $2 + 1$?) Typically, knocking down is taught in relation to "adding on" (making one more) as explained under **Something First**.

The answers to **One Less** will vary, but they will be something like the following:

1. Four - our - or
2. Limb
3. Tred - red
4. Plan - ant
5. Tee or ten
6. Ale
7. Sale - ale
8. Cut or Ute
9. Cod - or
10. Ran - in

One More / Less (p. 91)

A worksheet on adding on and making one less for two- and three-digit numbers. For different numbers, use the blank One More / Less on page 92.

Optical Illusions (p. 304)

Two cases of two lines each where one line looks longer than the other line but is not.

Children are to measure the lines on **Optical Illusions** and, hopefully conclude that some things need to be measured, that not everything can be simply "eyeballed."

Paths (p. 303)

A problem solving activity on measuring in centimeters.

For problem 1, the shortest path is 20 centimeters. For problem 2, the shortest path is 25 centimeters.

People Pictures (p. 11)

A set of manipulatives for sorting and ordering on the basis of sex (male or female), age (child or adult), build (skinny or fat) and dress (light or dark). A classroom set of them would consist of one set per child.

The **People Pictures** are used to assist with the development of the concept of number and give meaning to what is done with sets. To this end, they are sorted and ordered as explained under **Attribute Shapes** and used with **Going to the Park**.

Pick-a-Pair

A game for two to four on making one-to-one correspondences.

Materials:

One set of Pick-a-Pair cards (pp. 29-31) per group.

Directions:

To begin, each player draws a card. The player drawing the card exhibiting the greatest number of objects shuffles the cards and spreads them out face down to make a four-by-six array. The player to the left of the player making the array plays first. The play rotates clockwise.

To play, a player draws two cards from the array. If the cards exhibit the same number of objects, the player keeps the cards and draws again, and so on. If the cards do not exhibit the same number of objects, the player returns the cards to the array exactly as before, and the play goes to the next player.

The player with the most cards after all the cards have been paired wins.

Place Value Rummy

A game for two to four on base ten numeration.

Materials:

One deck of Place Value Rummy cards (pp. 111-117) per group.

Directions:

The same as for Addition Facts Rummy except without runs.

An example of a spread for the Place Value Rummy cards would be the

"3," "three ones," and picture of three units. An example of a book the circled three ones as well.

Puppies (p. 135)

A worksheet on the addition facts. With different numbers, as with the blank Puppies on page 136, a worksheet on any of the basic facts.

Racer (p. 221)

A worksheet on mixed facts. For different facts, use the blank Racer on page 222.

Ring Around the Rosy

A game for two to four on the "easy" multiplication facts -- the facts for which both factors are less than or equal to six. With different numbers, as with the blank Ring Around the Rosy gameboard on page 190, a game on any of the basic facts.

Materials:

Two dice and one Ring Around the Rosy gameboard (p. 189) per group. One marker per player.

Directions:

To begin, the players put their markers on the Start circle on the gameboard. Then each player rolls the dice. The player rolling the highest total plays first. The play rotates clockwise.

To play, a player rolls the dice and forms the product of the numbers showing. Then, if the product appears on the NEXT CLOSEST FLOWER on the gameboard, the player moves to that flower. Otherwise, the player stays where he or she is. The play then goes to the next player.

The first player to circuit the flowers on the gameboard wins.

Rod Patterns (pp. 44-45)

Self-directed activities on constructing patterns with colored rods.

A rod pattern is used as a referent for ordinal numbers. Relative to a rod pattern, children can be asked to point to the "first" rod, the "second" rod, the "third" rod, and so on.

Rulers (p. 302)

Six rulers, of which two are in inches, two in only centimeters, and two in centimeters and millimeters.

Safari

A counting game for two to four similar to Collect-a-Shape.

Materials:

One die and one Safari gameboard (pp. 56-57) per group. One marker and one set of Safari animals (p. 58) per player.

Directions:

In preparation, all the animals are combined and put in a tray of some sort so where they can be gotten to easily.

To begin, the players put their markers on the Start foot on the gameboard. Then each player rolls the die. The player rolling the highest number plays first. The play rotates clockwise.

To play, a player rolls the die and moves his or her marker along the feet and water holes on the gameboard in accordance with the number showing on the die. If the player lands on a foot, the play goes to the next player. If in a water hole, the player takes the animal for that water hole, and the play goes to the next player.

The first player to get five different animals wins.

Sameness Trains (pp. 17, 22)

Self-directed activities for ordering the Attribute Shapes on pages 14 and 15 and the Word cards on pages 18-20.

A "sameness train" is a string of "cars" for which the cars next to one another are the same in a predetermined number of ways. The objective in having children make a sameness train is to have them exercise logical thinking. (e.g., "How are the cars the same? How are they not the same?") And once made, a sameness train can be used as a referent for ordinal numbers.

(e.g., "Which is the first car? The second car? The third car?")

Some possibilities for the next car in the one-sameness train for the At tribute Shapes would be the small red circle (only the same size) or the large blue square (only the same color). For the Word cards, the word had (only the same number of letters) or the word they (only the same first letter). And some possibilities for the next car in the two-sameness train for the Attribute Shapes would be the small yellow rectangle (only the same size and shape) or the small red square (only the same size and color). For the Word cards, the word and (only the same first letter and number of letters) or the word the (only the same last letter and number of letters).

Seesaw (p. 157)

A related facts activity on addition and subtraction.

Seesaw is a good follow-up to working with a balance beam. (See Balance Beam Cutout.) For different numbers, use the blank Seesaw on page 158.

Sequence Cards (pp. 40-43)

Four sets of six cards each on ordination.

The cards in each set of Sequence cards are to be ordered from start to finish in terms of the event they depict. Once ordered, they may be used as a referent for giving meaning to the ordinal numbers "first," "second," "third," "fourth," "fifth," and "sixth."

Shape Rummy

A game for two to four on shape recognition.

Materials:

One deck of Shape Rummy cards (pp. 283-289) per group.

Directions:

The same as for Addition Facts Rummy except without runs.

An example of a spread for the Shape Rummy cards would be the three segments. An example of a book the word SEGMENT as well.

Shooting Gallery

A game for two on the addition facts. With different numbers, as with the blank Shooting Gallery gameboard on page 145, a game on any of the basic facts.

Materials:

One die and one Shooting Gallery gameboard (p. 144) per group. Six markers per player.

Directions:

To begin, the players decide on who gets which of the two sets of six targets each on the gameboard. Then each player rolls the die. The player rolling the higher number plays first.

To play, the players take turns rolling the die and "shooting" each other's targets by putting a marker on the target of the same number as the number showing on the die.

The first player to shoot all of the other player's targets wins.

Variation:

Have the players play with a consumable copy of the gameboard and color their own targets instead of shooting each other's targets. Then make the winner the first player to color all of his or her targets.

Shopping Cards (pp. 266-268)

Twenty-four "shopping" cards with which to present exercises on problem solving.

The **Shopping cards** are used to conduct problem solving exercises on the four operations as illustrated below:

Addition:

"What could you buy for 10 cents?

"What **three** items could you buy for 25 cents?

"What could **Motley Crab Adder** combine if he had 15 cents?

"What **four** items could **Motley Crab Adder** combine if he had $1.00?"

Subtraction:

"How much change would you get if you gave a store clerk a nickel for the grapes? Fifty cents for the oranges?

"What would **The Scruffy Twin Subtractors** do if you gave them a quarter for the taco plate? One dollar for the donuts"?

Multiplication:

"How much for three strip steaks? Five hamburger plates?

"How much money would **Sir Crab Multiplier** need if he wanted to combine four boxes of cherries? Six cheese wedges?"

Division:

"How much for one hotdog? One apple?

"What would **The Impeccable Twin Dividers** do with 40 cents if they wanted just one orange? With $1.00 if they wanted just one nectarine?"

Sir Crab Multiplier (p. 259)

The personification of multiplication as combining "neatly" (by twos, threes, fours, and so on).

Sir Crab Multiplier is used to help children understand arithmetic word problems as explained under **Motley Crab Adder**. (For specific instances of his use, see **Dot Paper** and **Shopping Cards**.) Like Motley, he can be made into an attractive item for the bulletin board or into an attractive poster. And also like Motley, he is independent of computational skill and can therefore be talked about to even the youngest of children to help them see the mathematics implicit in the world around them.

An introduction to Sir Crab would go something like this:

"Meet **Sir Crab Multiplier**, a strange character indeed. He always combines things 'neatly' (by twos, threes, fours, and so on). That's just the way he is. No one knows why.

"People have talked to him about this and have told him that regardless of all the 'neat' brick walls and things he's made, he's driving them crazy with the way he always combines things neatly. They have even suggested that the next time he wants to combine some things neatly, he stick his head in a bucket instead. But he just blows them a kiss and goes on combining things neatly.

"And marvel at how neatly he's dressed, just what you'd expect from someone who combines things neatly. No wonder he's addressed as 'Sir.'

"And look at all the things he can do!" (Mime some of the following:

> Making a rug or a beaded belt
>
> Picking four leaf clovers
>
> Freezing water in ice cube trays
>
> Stacking wood or timber
>
> Exchanging dollars for pennies, nickles, dimes, or quarters a dollar at a time

"Can you show me these things he can do?" (Have children mime some of the following:

> Soaking up water with a sponge
>
> Blowing up a balloon
>
> Buying cartons of eggs
>
> Collecting animals for Noah's ark
>
> Laying floor or roof tiles

Ski Slope

A game for two to four on division of whole numbers.

Materials:

One Ski Slope gameboard (pp. 208-209) and one set of Division cards (pp. 210-211) per group. One marker per player.

Directions:

To begin, the players put their markers on the Start space on the gameboard. Then each player draws a Division card. The player drawing the card with the highest remainder plays first. The play rotates clockwise.

To play, a player draws a Division card, figures the remainder for the card, and skis down the slope on the gameboard as many spaces as the remainder. The play then goes to the next player.

Once a player reaches the lift at the bottom of the gameboard, the player takes the lift to the Start space and skis down the slope as before.

The first player to ski down the slope three times wins.

Variation:

Put the following numbers on the gameboard in the blank spaces or in the boxes inside the spaces with writing in them, one number per space or box: 12, 15, 21, 28, 30, 35, 42, 45, 54, 56, 63, and 72. Then, instead of having a player draw a Division card, have a player roll two dice and ski down the slope in accordance with the remainder from dividing the number the player is on by the total on the dice.

Smoke Rings (p. 73)

A numeration activity on the number names from one to five, inclusive.

Snail Trail

A game for two to four on the addition and subtraction facts.

Materials:

Three dice per group. One Snail Trail gameboard (p. 166) and 15 (or 18 i playing the variation) markers per player.

Directions:

To begin, each player rolls the dice. The player rolling the highest tota plays first. The play rotates clockwise.

To play, a player rolls the dice and strives to make one of the numbers on his or her gameboard by adding or subtracting the numbers showing on the dice. The rule is to use each of the numbers on the dice once and only once. To illustrate, if a player rolls a 2, 3, and 4, the player could make the 1 $(2 + 3 - 4)$, 3 $(2 + 4 - 3)$, 5 $(3 + 4 - 2)$, or 9 $(2 + 3 + 4)$.

Once a player makes a number, the player covers the number with a marker, and the play goes to the next player. If a player cannot make a number still uncovered, the player misses that turn.

The first player to cover 15 numbers wins.

Variation:

Allow for the numbers showing on the dice to be multiplied and divided as well as illustrated in the directions for Moon Walk and Trizo. Then make the winner the first player to cover all 18 numbers on his or her gameboard.

Something First (p. 173)

An integration of math and language arts on "adding on" (making one more).

Adding on is a skill taught to children in teaching them the addition facts. First, they are taught the "doubles" $(1 + 1 = 2, 2 + 2 = 4, 3 + 3 = 6, \ldots)$. Then they are shown how knowing how to "add on" gives them a great many more addition facts. (For example, "If $2 + 2 = 4$, what's $2 + 3$?") Typically, adding on is taught in relation to "knocking down" (making one less) as explained under **One Less**.

The answers to **Something First** will vary, but they will be something like the following:

1. Fit - slit
2. Care - share
3. Crow - throw
4. Train - strain
5. Ran - than
6. Lease - please
7. This
8. Height - freight
9. Bin - shin
10. Sand - grand
11. Core - score
12. Dart - start
13. Bone - stone
14. Sat - flat
15. Those

Speedy Operator (p. 232)

A self-correcting exercise on whole number arithmetic. For different problems, use the blank Speedy Operator on page 233.

A child is to start with the number in the Starter Box and work his or her way around the track from racer to racer by following the instructions on the flags and writing the answers in the racers. The rule is to use the answer for one racer with the instruction in the flag for the next racer to get the answer for that racer. Thus the answers for the first three racers would be 20 (10 × 2), 15 (20 − 5), and 45 (15 × 3).

If a child's answer for the last racer does not agree with the answer in the Winner's Circle, the child is to start again and find the racer that "crashed."

Spider and Fly

A game for two to four on whole number arithmetic. For different problems, use the blank Spider and Fly gameboard on page 235.

Materials:

For each group, a die and a Spider and Fly gameboard (p. 234). For each player, a marker and a pen or pencil and some scrap paper.

Directions:

To begin, the players put their markers on the spider on the gameboard. Then each player rolls the die. The player rolling the highest number plays first. The play rotates clockwise.

To play, a player rolls the die and moves his or her marker along the numbered spaces on the gameboard as many spaces as the number showing on the die. The player then works the problem in the space he or she lands on and moves to the space of the same number as the answer to the problem. To illustrate, if a player is on the spider and rolls a four, the player moves four spaces to space four, works the problem there, namely 4 + 7, and then moves to space 11 in keeping with the answer to the problem. The play then goes to the next player.

The first player to reach the fly on the gameboard wins.

The answers to the problems for Spider and Fly are given below.

1. 4	11. 12	21. 18	31. 43	41. 42
2. 6	12. 9	22. 25	32. 28	42. 44
3. 5	13. 22	23. 36	33. 27	43. 48
4. 11	14. 24	24. 31	34. 42	44. 40
5. 3	15. 18	25. 30	35. 36	45. 47
6. 16	16. 9	26. 36	36. 39	46. 30
7. 10	17. 12	27. 21	37. 47	47. 48
8. 9	18. 22	28. 26	38. 38	48. 50
9. 15	19. 25	29. 33	39. 30	49. 50
10. 9	20. 22	30. 25	40. 41	50. Blank

Spinners (pp. 2-3)

Spinners with 3, 4, 5, 6, 8, or 10 sides. For different numbers on the spinners, use the blank spinners on pages 4 and 5.

The first use of the spinners is as a substitute for a die or dice in a game. The second is as the makings of a game in their own right as will be explained after the following paragraph.

To make the spinners, pin them to boards or pencils with thumbtacks as in Figures 20 and 21, respectively.

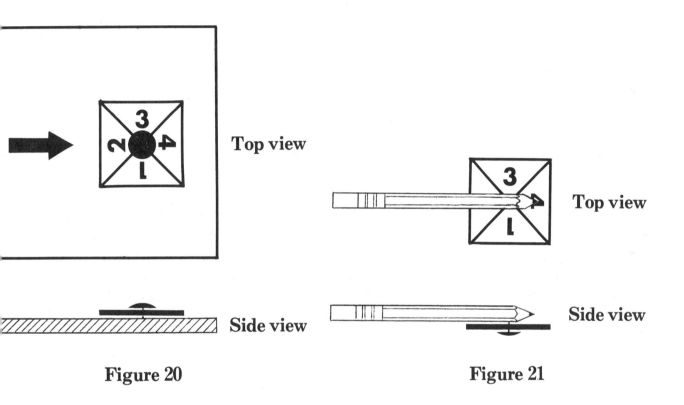

Figure 20	Figure 21

Note that in either case it is the face of the spinner that spins, and that in the first case the indicator arrow is drawn on the board, whereas in the second case the pencil itself is the indicator. To lessen the tendency of the pencil or "wand" model to yield the same number, hold the pencil horizontally.

To make a game out of the spinners, set a goal such as 21 and have children try to reach the goal by spinning two or three spinners and adding, subtracting, multiplying, or dividing the numbers indicated in any way possible. Have them record their results and keep a running total, and make the winner the first one to reach the goal.

Staircase (p. 140)

A variation of Boat.

Stepping Stones

A game for two to four on the multiplication and division facts. With different numbers, as with the blank Stepping Stones gameboard on pages 218 and 219, a game on any of the basic facts.

Materials:

One die and one Stepping Stones gameboard (pp. 216-217) per group. One marker per player.

Directions:

To begin, the players put their markers on the Start side of the river on the gameboard. Then each player rolls the die. The player rolling the highest number plays first. The play rotates clockwise.

To play, a player rolls the die and moves his or her marker across the river along the path he or she is on to the nearest stepping stone or island of the same number as the number showing on the die. If the player lands on a stepping stone or island with a log on it, he or she crosses the log to the next stepping stone or island. If on a stepping stone with an alligator present, he or she moves two stepping stones or islands back. The play then goes to the next player.

The first player to cross the river wins.

Subtraction in Different Lands

An activity on the concepts of base and place value in relation to subtraction of whole numbers.

As explained under **Addition in Different Lands**, the objective of **Subtraction in Different Lands** is not to teach children how to subtract in different bases, but rather when and how to exchange when subtracting. To this end, it illustrates that base is only a number that tells you how much of one thing you get for another and place value only a way of showing how much of whatever you end up with.

Materials:

For each child, a set of multi-base blocks or a four-color assortment of counters and one of either of the two **Subtraction in Different Lands worksheets** (pp. 105-106).

Procedure:

For best results, precede with **Clear It** and **Break a Cube.** (See **Bank It or Clear It** and **Build a Cube or Break a Cube.**) Then present the worksheet in a way that covers the following points:

For each problem, note the land it is in and illustrate the minuend (the top number or "starting amount") with the blocks (or counter equivalent of the blocks) for that land by illustrating a numeral in the unit column with that many units, a numeral in the long column with that many longs, a numeral in the flat column with that many flats, and a numeral in the cube column with that many cubes.

Once the minuend is illustrated with the blocks, separate the blocks into two piles such that the blocks in one of the piles illustrate the subtrahend (the bottom number or "amount to be subtracted"). To separate them, you may have to exchange a long for some units, a flat for some longs, or a cube for some flats. (As with **Clear It** and **Break a Cube,** there is no going to jail here.)

What you end up with in the other pile will be the difference. Record it beneath the subtrahend being careful to record the digits for it in the proper columns.

The answers for the worksheets are as follows:

Page 105:

a. 213
b. 1333
c. 204
d. 3404
e. 1546

Page 106:

a. 1117
b. 138
c. 2638
d. 919
e. 160

Subtraction Magic (p. 229)

A magic show on subtraction.

The explanation is beyond the aim of this book, but if children put any 1-, 2-, or 3-digit numbers in the outermost corners of either of the large squares on **Subtraction Magic** and keep subtracting the smaller number, they will always arrive at identical numbers somewhere in the process. In particular, they will do so for numbers of special significance to them, numbers such as their ages in months, their heights in inches, or the first three digits of their telephone numbers.

An example of **Subtraction Magic** for the numbers 483, 70, 41, and 179 is given below.

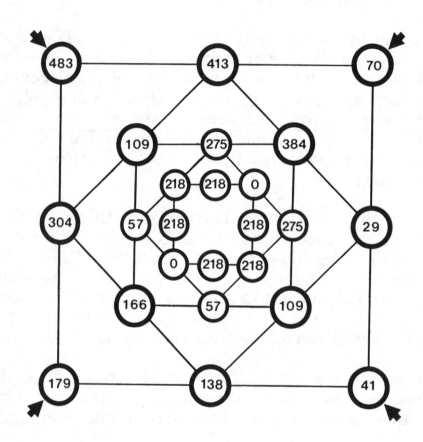

Super Bowl

A game for two on whole number arithmetic, the significance, of which, is the same as that for Doll.

Materials:

For each group, a Super Bowl gameboard (p. 249), a set of Super Bowl cards (p. 250), a deck of ASMD cards (pp. 251-254), and an ASMD key (p. 255). For each player, a marker, an ASMD scorecard (p. 256), and a pen or pencil and some scrap paper.

Directions:

Super Bowl is a take off on football.

In preparation, the Super Bowl cards and ASMD cards are shuffled

ogether and laid face down in the center of the gameboard. Then the players
decide on who gets which end zone and who is to play first.

To play, a player puts his or her marker on the gameboard on the 20-yard
line nearest his or her end zone and draws a card. If it is a Super Bowl card, the
player does what it says. If it is an ASMD card, the player works one of the
problems on the card and checks his or her answer to the problem using the
ASMD key. If incorrect, the player records a "miss" in the appropriate space
on his or her ASMD scorecard, removes his or her marker from the gameboard,
and the play goes to the other player. If correct, the player records a "hit"
instead, moves his or her marker 10 yards in the direction of the opposition's
end zone, draws another card, and proceeds as before.

The objective is to maintain an unbroken string of correct answers so as to
reach the opposition's end zone and score a touchdown. A touchdown is worth
six points. Once a touchdown is scored, the play goes to the other player.

The player with the most points after all the cards have been drawn wins.

To speed up the game, have players move 20 yards each time they work a
problem correctly.

Variation:

Allow an extra point after a touchdown for the correct answer to a second
problem from the last card drawn.

Take the Children Home (p. 179)

A worksheet on the multiplication facts. With different numbers, as with
the blank Take the Children Home on page 180, a worksheet on any of the basic
facts.

Tangram Pieces (p. 315)

Two sets of Tangram Pieces. A classroom set of Tangram Pieces would
consist of one set per child.

The **Tangram Pieces** are seven geometric figures which, to begin with,
make a square, but, when rearranged, make birds, cats, houses, and many
other shapes. They are used primarily to develop the concept of area as with
the **Tangram Task Cards** on pages 316-323.

Exploring area with Tangram Pieces

Tangram Task Cards (pp. 316-323)

Twelve task cards for the Tangram Pieces on page 315.

The answers for the task cards requiring answers are given below.

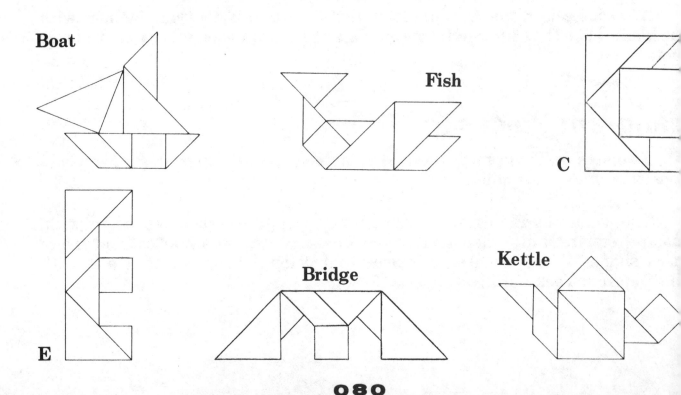

Boat

Fish

C

E

Bridge

Kettle

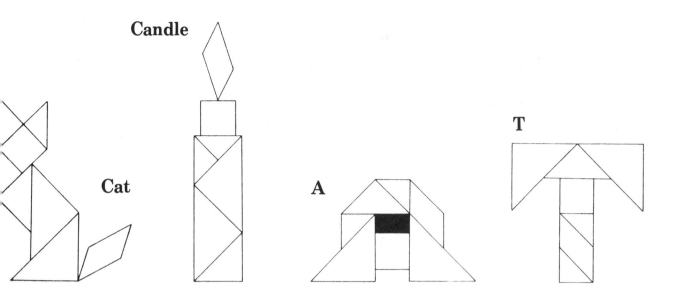

Candle

Cat

A

T

Comparing:

The figures are the same size because each of them can be covered with the same **Tangram Pieces** as shown below:

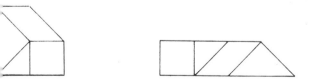

Area:

The trapezoid and the pentagon have the same area because each of them can be covered with the same **Tangram Pieces** as shown below:

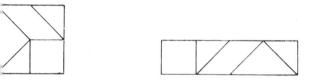

In contrast, the rectangle can be covered using only two of the same **Tangram Pieces** as shown below:

Teeter Totter (p. 212)

A related facts activity on multiplication and division.

Teeter Totter is a good follow-up to working with a balance beam. (See Balance Beam Cutout.) For different numbers, use the blank Teeter Totter on page 213.

Telephone

A counting game for two to four on the numerals 0, 1, 2, . . . , 9.

Materials:

One die and one Telephone gameboard (pp. 62-63) per group. One marker and two each of the numerals 0, 1, 2, . . . , 9 on page 64 per player.

Directions:

In preparation, all the numerals are combined and put in a tray of some sort to where they can be gotten to easily.

To begin, the players put their markers on the Start arrow on the gameboard. Then each player rolls the die. The player rolling the highest number plays first. The play rotates clockwise.

To play, a player rolls the die and moves his or her marker along the arrow and numerals on the gameboard in accordance with the number showing on the die. If the player lands on the arrow, the play goes to the next player. If on a numeral, the player takes a numeral like it up to two such numerals, and the play goes to the next player.

The objective is to make one of the telephone numbers on the gameboard with the numerals. The first player to do so wins.

Variation:

Make the winner the first player to make their own telephone number.

Ten Land Task Cards (pp. 107-108)

Two task cards on base ten numeration and addition and subtraction of whole numbers in terms of base ten blocks.

The Ten Land task cards are used to give meaning to base ten numeration and addition and subtraction of whole numbers. They are easily worked it preceded with **Bank It or Clear It, Build a Cube or Break a Cube, Addition in Different Lands,** and **Subtraction in Different Lands.** For additional task cards like them, use the small multi-base block cutouts on pages 96-98.

The answers for the task cards are as follows:

Card A:

1. 2535

Card B:

1. 1618
2. 4153
3. 917

The Great Legalizer (p. 261)

The personification of the need in computing with whole numbers to regroup.

The Great Legalizer is used to help children regroup properly as explained below and illustrated for **Bank It, Build a Cube,** and **Addition in Different Lands.** In that **The Great Legalizer** represents one of the six processes fundamental to basic mathematics (as explained under **Motley Crab Adder**), he is worthy of display as on a bulletin board or classroom wall and is worth talking about to children. Since he is independent of computational skill, he can be talked about to even the youngest of children.

An introduction to **The Great Legalizer** would go something like this:

> "Meet **The Great Legalizer,** the keeper of the law in the different lands. He always blows his whistle to remind someone with too many of one thing alike to exchange them for one of the next larger thing. You see, the law in two land is 'Never, never get caught with two or more things alike else GO TO JAIL!' The law in three land is 'Never, never get caught with three or more things alike else GO TO JAIL!' The law in four land is 'Never, never get caught with four or more things alike else GO TO JAIL!'" And so on.

> "So what would **The Great Legalizer** do if he saw someone with, say, two things alike in two land?" (Blow his whistle to remind them to exchange them for one of the next larger thing.) "Three things alike in three land?" (Blow his whistle to remind them to exchange them for one of the next larger thing.) "Fifteen things alike in ten land?" (Blow his whistle to remind them to exchange ten of them for one of the next larger thing.) And so on.

"So you see, **The Great Legalizer** is a big help to **Motley Crab Adder** and **Sir Crab Multiplier** who are always ending up with too many of one thing alike. He can also be a big help to you!"

Blowing the whistle on Motley Crab Adder to remind him to "not go to jail"

The introduction is greatly enhanced by having children use multi-base blocks or different colored counters to actually make the above exchanges (and additional exchanges).

The Impeccable Twin Dividers (p. 260)

The personification of division as separating "neatly" (into twos, threes, fours, and so on).

The **Impeccable Twin Dividers** are used to help children understand arithmetic word problems as explained under **Motley Crab Adder.** (For specific instances of their use, see **Dot Paper** and **Shopping Cards.**) Like Motley, they can be made into an attractive item for the bulletin board or into an attractive poster. And also like Motley, they are independent of computational skill and can therefore be talked about to even the youngest of children to help them see the mathematics implicit in the world around them.

An introduction to the Impeccable Twins would go something like this:

"Meet **The Impeccable Twin Dividers**, strange characters indeed. They always separate things 'neatly' (into twos, threes, fours, and so on). That's just the way they are. No one knows why.

"People have talked to them about their style and have tried to get them to relax and stop all this separating neatly. They have even gone so far as to suggest that if they do it just one more time, they will have them committed. But **The Impeccable Twin Dividers** just smooth their clothes and go on separating things neatly.

"And speaking of their clothes, look at how neatly they wear them, just what you'd expect of a twosome who separate neatly. Small wonder that part of their name is 'impeccable.'

"And look at all the things they can do!" (Mime some of the following:

> Unraveling knitting
>
> Mowing a lawn in strips
>
> Picking teams
>
> Sharing money equitably
>
> Tearing down a brick wall a row at a time

"Can you show me these things they can do?" (Have children mime some of the following:

> Folding socks
>
> Cleaning venetian blinds
>
> Bailing out a boat
>
> Dealing out a deck of cards
>
> Taking handfuls of things from a container like nuts from a bowl or chocolates from a box

The Magnificent Equalizer (p. 262)

The personification of the need in working with fractions to make things "different yet the same" -- different in appearance, yet the same in quantity.

The **Magnificent Equalizer** is used to help children understand equivalent fractions as explained below. As a representative of one of the six processes fundamental to basic mathematics as explained under **Motley Crab Adder,** he is worthy of display (as on a bulletin board or classroom wall) and is worth talking about to children. Since he is independent of computational skill, he can be talked about to even the youngest of children.

An introduction to **The Magnificent Equalizer** would go something like this:

> "Meet **The Magnificent Equalizer,** a truly wonderous person. He always makes things 'different yet the same' -- different in appearance, yet the same in quantity. That's just the way he is. No one knows why.

> "People have talked to him about this and have told him that he is weird because of it. And many of them, believing that some things are better left alone, have taken to hiding things from him. But he, believing that everything can stand a little change, just goes on making things different yet the same. This shows in the equal sign on his chest and the equal weights, however sectioned, on the ends of his barbell.

> "You see, he knows that most of the people criticizing him are just jealous, that they wish that they too could make things different yet the same. And small wonder. If they could, just imagine some of the things they could do.

> "With only a click of their fingers, they could get the lawn mowed without all the mowing (that is, a different looking lawn even though the same lawn). They could get the garbage carried out without all the carrying (that is, a different location for the garbage even though the same garbage). They could even get rid of some of their bad habits without all the self-discipline (that is, a different set of behaviors even though the same persons).

> "But enough of the wishful thinking. Let me show you some of the more down-to-earth things he can make different yet the same, things which you, too, can make

different yet the same.

"He can take a simple piece of paper and use it to turn 1 / 2 into 2 / 4. And watch! You can do it too." (Have each child take a rectangular piece of paper, fold it in half, open it up, and color half of it. Then have all the children fold their papers back as they were, fold them in half again, and open them up. Each of them will see 1 / 2 = 2 / 4, that is, a different way of viewing the same coloring.)

"And using the same piece of paper, he can turn 3 / 4 into 6 / 8, but, then, so can you." (Have each child color some more of his or her paper to where 3 / 4 of it is colored. Then have all the children fold their papers back as they were after the last folding, fold them in half again, and open them up. Each of them will see 3 / 4 = 6 / 8, thus, one again, a different way of viewing the same coloring.)

"And that's not all. He can even use this piece of paper to change 7 / 8 into 14 / 16. Can you show me what he would do?" (Have the children proceed as before.)

"What do you think he would turn 15 / 16 into? How could you prove your answer?"

The Scruffy Twin Subtractors (p. 258)

The personification of subtraction as separating.

The Scruffy Twin Subtractors are used to help children understand arithmetic word problems as explained under **Motley Crab Adder.** (For specific instances of their use, see **Dot Paper** and **Shopping Cards.**) Like Motley, they can be made into an attractive item for the bulletin board or into an attractive poster. And also like Motley, they are independent of computational skill and can therefore be talked about to even the youngest of children to help them see the mathematics implicit in the world around them.

An introduction to the Scruffy Twins would go something like this:

"Meet **The Scruffy Twin Subtractors,** strange characters indeed. They always separate things. That's just the way they are. No one knows why.

"People have talked to them about this and have tried to get them to change and stop all this separating. They have even said that they'll put them in straight jackets if they separate anything again. But **The Scruffy Twin Subtractors** just shrug their shoulders and go on separating.

"Also, when they separate things, they separate them in just any old way. This shows in their tattered clothes and is why part of their name is 'scruffy.'

"And look at all the things they can do!" (Mime some of the following:

> Trimming fat off a steak
>
> Whittling a stick
>
> Mowing the lawn
>
> Robbing a piggy bank
>
> Taking a bite out of something like a pear or an apple

"Can you show me these things they can do?" (Have children mime some of the following:

> Scaling a fish
>
> Peeling a banana, orange, potato, or the like
>
> Spending part of a dollar
>
> Clipping or filing a toenail or a fingernail
>
> Ripping a page out of a phone book

"Separating" paper just like The Scruffy Twin Subtractors

Drawing The Scruffy Twin Subtractors

Cutting out The Scruffy Twin Subtractors

Voila! The Scruffy Twin Subtractors

Three-Way Sort Task Card (p. 24)

A blank task card for classifying relative to three attributes.

The **Three-Way Sort task card** is used in the manner explained under **Two-Way Sort Task Card.**

Some examples of some classifications for the **Three-Way Sort task card** are given below.

People Pictures:

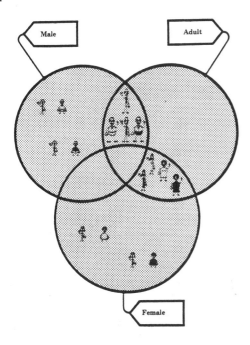

Attribute Shapes:

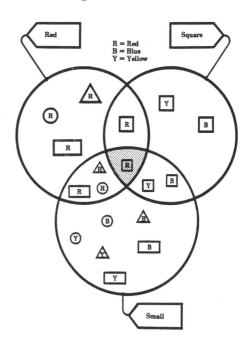

Word Cards:

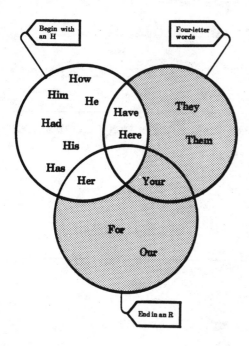

The shading in the example for the **People Pictures** illustrates the **union** o the set of males, the set of adults, and the set of females, that is, the set of males adults, **or** females. The shading in the example for the **Attribute Shape** illustrates the **intersection** of the set of red figures, the set of square figures, an the set of small figures, that is, the set of red, square, **and** small figures. And th shading in the example for the **Word cards** illustrates the **complement** of the se of words that begin with an **H** relative to the set of four-letter words and the set o words that end in an **R**, that is, the set of four-letter words that do **not** begin wit an **H** or words that end in an **R** that do **not** begin with an **H**.

Toys (p. 153)

A worksheet on the subtraction facts. With different numbers, as with th blank Toys on page 154, a worksheet on any of the basic facts.

Tree (p. 187)

A variation of Caboose.

Trizo

A game for two to four on the addition, subtraction, multiplication, an division facts.

Materials:

For each group, a Trizo gameboard (p. 224) and three dice. For each player, approximately 10 markers of a particular color.

Directions:

Trizo is a take off on the game of tic-tac-toe.

To begin, each player rolls the dice. The player rolling the highest total plays first. The play rotates clockwise.

To play, a player rolls the dice and strives to make one of the numbers on the gameboard by adding, subtracting, multiplying, or dividing the numbers showing on the dice. The rule is to use each of the numbers on the dice once and only once. To illustrate, if a player rolls a 3, 3, and 4, the player could make the 2 $(3 + 3 - 4)$, 3 $(3 \times (4 - 3))$, 4 $(4 + 3 - 3)$, 5 $(4 + 3 \div 3)$, 9 $(3 \times 4 - 3)$, 10 $(3 + 3 + 4)$, 13 $(3 \times 3 + 4)$, 15 $(3 \times 4 + 3)$, or 36 $(3 \times 3 \times 4)$.

Once a player makes a number, the player "captures" the number by covering it with a marker, and the play goes to the next player. If a player cannot make a number not already captured, the player misses that turn.

The first player to capture three horizontally, vertically, or diagonally adjacent numbers wins. Alternatively, the first player to capture two numbers which, in combination with a free space, are horizontally, vertically, or diagonally adjacent wins.

The reason for the free spaces being where they are on the gameboard is to encourage the use of multiplication.

Two-Way Sort Task Card (p. 23)

A blank task card for classifying relative to two attributes.

The Two-Way Sort task card is used to classify the People Pictures, Attribute Shapes, and Word cards. In that the task card is nearly non-verbal, it can be used by very young children. Its value is that it exercises logical thinking and gives meaning to the union, intersection, and complement of sets.

To prepare the task card for use, fill in the labels for the "loops" on the task card as follows: If the task card is to be used with the People Pictures, fill in the labels with words like **male** and **adult**. If with the Attribute Shapes, with words like **red** and **square**. And if with the Word cards, with phrases like "Begin with an H" and "Four-letter words." The labels then specify what is to go inside the loops.

Some examples of some classifications for the task card are given below.

People Pictures:

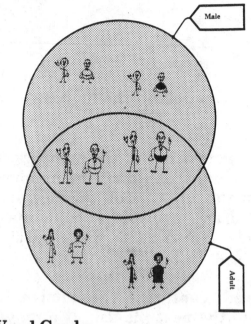

Attribute Shapes:

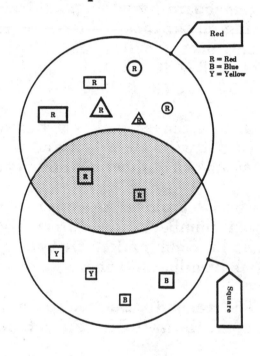

Word Cards:

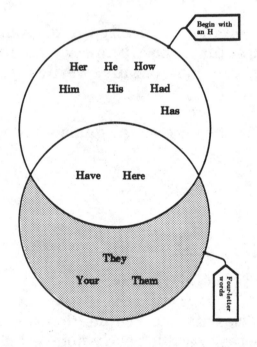

The shading in the example for the **People Pictures** illustrates the **union** of the set of males and the set of adults, that is, the set of males **or** adults. The shading in the example for the **Attribute Shapes** illustrates the **intersection** of the set of red figures and the set of square figures, that is, the set of red **and** square figures. And the shading in the example for the **Word cards** illustrates the **complement** of the set of words that begin with an **H** relative to the set of four-letter words, that is, the set of four-letter words that do **not** begin with an **H**.

Vertices, Faces, and Edges (pp. 291-293)

Three exercises on the vertices, faces, and edges of networks: two-dimensional, "all hooked up" figures as explained for problem 1 on page 291.

The exercises on the vertices, faces, and edges of networks are of special interest in that they integrate arithmetic, algebra, and geometry: arithmetic in the computing of the sums and differences involved, algebra in the use of the variables V = **number of vertices**, F = **number of faces**, and E = **number of edges**, and geometry in the rule that surfaces, that $V + F - E = 1$ for the vertices, faces, and edges of any network. Thus for page 291, $V + F - E = 1$, because each of the figures consists of one network. For page 292, $V + F - E = 2$, because each of the figures consists of two networks. And for page 293, $V + F - E = 3$, because each of the figures consists of three networks.

The answers for the worksheets are as follows:

Page 291:

Figure	V	F	E	V + F − E
a	5	4	8	1
b	8	5	12	1
c	12	7	18	1
d	8	6	13	1
e	8	8	15	1
f	5	3	7	1
g	16	9	24	1
h	6	2	7	1
i	12	6	17	1
j	10	7	16	1

1. A two-dimensional figure which allows for getting from any point on the figure to any other point on the figure while staying on the figure

2. 549
3. 8001
4. 22,824,883
5. $V + F - E = 1$

Page 292:

Figure	V	F	E	V + F − E
a	16	6	20	2
b	16	5	19	2
c	14	4	16	2
d	15	7	20	2
e	13	2	13	2
f	10	7	15	2
g	8	2	8	2
h	6	3	7	2
i	15	4	17	2
j	9	3	10	2

1. Two

2. The number of networks is the same as $V + F - E$ for the networks.

3. a. One, except for i and j. For them, it is two.

   b. One

   c. Answers will vary, but nine for a name like Ima Silly

   d. Answers will vary, but four for a name like *Ima Silly*

Page 293:

Figure	V	F	E	V + F — E
a	13	4	14	3
b	13	1	11	3
c	8	6	11	3
d	10	1	8	3
e	19	8	24	3
f	21	6	24	3
g	8	4	9	3
h	8	6	11	3
i	13	3	13	3
j	14	5	16	3

1. Three

2. Just count networks. The number of networks is the same as **V + F — E** for the networks.

3. a. 4

   b. Answers will vary, but five for a drawing like the one below

Walk a Crooked Meter

A game for two to four on millimeters, centimeters, and decimeters.

Materials:

For each group, a Walk a Crooked Meter gameboard (p. 305), a "shaker box" -- an egg carton with measurements written in the bottom of it -- and a small object to shake in the shaker box. For each player, a marker. For the shaker box, use measurements like 2 cm (centimeters), 30 mm (millimeters), and 1 dm (decimeters) that equate with whole numbers of centimeters.

Directions:

To begin, the players put their markers on the Start circle on the gameboard. Then each player shakes the shaker box. The player "shaking" the longest measurement plays first. The play rotates clockwise.

To play, a player shakes the shaker box and moves his or her marker along the centimeter trail on the gameboard the number of centimeters indicated. The play then goes to the next player.

The first player to "walk the crooked meter," that is, to walk the 100 centimeters = 1 meter to the end of the trail wins.

Variation:

The same game except with a meter stick in place of the gameboard and a box of colored rods in place of the markers. The objective then becomes to be the first player to lay out a meter's worth of colored rods.

Watermelon

A game for two similar to Car Park on making one-to-one correspondences.

Materials:

One Watermelon gameboard (p. 27) and 16 "seeds" (p. 28) per player. Some envelopes or match boxes for the seeds.

Directions:

In preparation, the seeds are packaged in varying amounts in the envelopes or match boxes.

To play, the players take turns taking envelopes or match boxes and putting the seeds therein on their gameboards in the spaces provided, one seed per space.

Playing Watermelon

The first player to fill in all the spaces on his or her gameboard wins.

Variation:

The same game except with egg cartons in place of the gameboard an[d] cardboard eggs in place of the seeds.

Witch (p. 181)

A worksheet on the multiplication facts. With different numbers, as with the blank Witch on page 182, a worksheet on any of the basic facts.

Wobble Town

An activity on constructing polyhedra and discovering the "fundamental principle of rigidness," the fact that the triangle is the only rigid polygon.

Materials:

For **Wobble Town**, approximately 200 straws in various lengths and 350 pipe cleaners in 2-inch lengths. For the workers, one **Wobble Town worksheet** (p. 290) per worker.

To supervise the building of **Wobble Town**, have the workers attach the straws with the pipe cleaners as shown below.

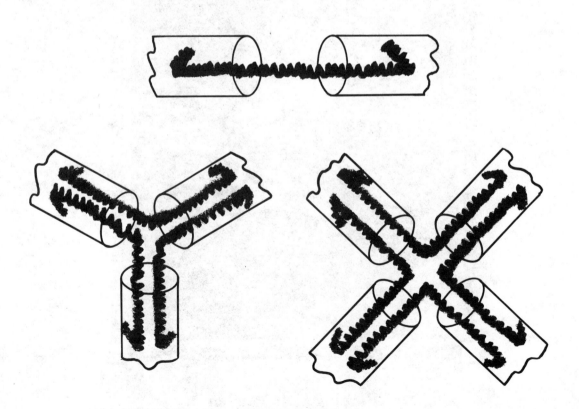

Be certain to insist on the small bend at the ends of the pipe cleaners. Without it, the straws will slip off the pipe cleaners.

Once the town is built, it can be used as a referent for a variety of lessons:

Have children draw some of the "buildings" that they might better understand the three-dimensional drawings on the **Wobble Town worksheet**.

Have children compare buildings B, C, G, I, and K to the remaining buildings and note the "specialness" of the former, the fact that each of them is made up of faces of the same size and shape. These buildings are the five and only five regular polyhedra. Their geometric names are as follows: B is called a **regular**

Making polyhedra with straws and pipe cleaners

icosahedron, C a **regular octahedron**, G a **regular tetrahedron**, I a **regular dodecahedron**, and K a **cube**.

Have children compare the "wobbly" buildings, buildings such as E, J, and K, to the "non-wobbly" buildings, buildings such as B, C, and G, and discover the fundamental principle of rigidness, the fact that the triangle is the only rigid polygon. Then have them turn some of the "wobblies" into "non-wobblies" by triangulating the wobblies. Afterwards, consider taking them for a walk outdoors and directing their attention to the many strengthening uses triangles are put to as in bracing walls, trussing bridges, and supporting towers.

Have children count the vertices, faces, and edges of each building, record their findings in a table like the following, and calculate $V + F - E$ for the building for V = **number of vertices** (where the pipe cleaners are), F = **number of faces** (what could be colored if the buildings were covered with paper), and E = **number of edges** (the straws).

Building	V	F	E	V + F − E
A				
B				
C				

When completed, the table should appear as follows wherein it becomes apparent that $V + F - E = 2$ "always," a true statement for polyhedra like the ones in **Wobble Town**, that is, for polyhedra which, if made of clay, could be rolled into a ball instead of, say, a doughnut.

Building	V	F	E	V + F − E
A	8	6	12	2
B	12	20	30	2
C	6	8	12	2
D	5	5	8	2
E	10	7	15	2
F	6	5	9	2
G	4	4	6	2
H	8	6	12	2
I	20	12	30	2
J	12	8	18	2
K	8	6	12	2

As a follow-up, have children complete the exercises on vertices, faces, and edges on pages 291-293. The exercises illustrate that for "networks," two-dimensional figures as explained in the exercises, $V + F − E = 1$ -- a one instead of a two! -- the "price," so to speak, that polyhedra "pay" to be made into two-dimensional figures by having one of their faces removed that they might be "smoothed" out.

Word Cards (pp. 18-20)

A set of manipulatives for sorting and ordering on the basis of "beginning letter," "number of letters," "ending letter," and other attributes (e.g., "number of vowels," "words with a repeating letter," and "words which refer to people"). Also, a list of 45 of the 100 most commonly used words in writing, thus a medium for integrating math and language arts. A classroom set of them would consist of one set per child.

The Word cards are used to assist with the development of the concept of number and give meaning to what is done with sets. To this end, they are sorted and ordered as explained under Attribute Shapes and used with Gimmi. For a specific instance of ordering them, see page 22.)

World Series

A game for two on whole number arithmetic, the significance, of which, is the same as that for Doll.

Materials:

For each group, a World Series gameboard (p. 247), a set of World Series cards (p. 248), a deck of ASMD cards (pp. 251-254), and an ASMD key (p. 255). For each player, a marker, an ASMD scorecard (p. 256), and a pen or pencil and some scrap paper.

Directions:

World Series is a take off on baseball.

In preparation, the World Series cards and ASMD cards are shuffled together and laid face down in the center of the gameboard. Then the players decide on who is to play first.

To play, a player puts his or her marker on home plate on the gameboard and draws a card. If it is a World Series card, the player does what it says. If it is an ASMD card, the player works one of the problems on the card and checks his or her answer to the problem using the ASMD key. If incorrect, the player records a "miss" in the appropriate space on his or her ASMD scorecard, removes his or her marker from the gameboard, and the play goes to the other player. If correct, the player records a "hit" instead, moves his or her marker to first base, draws another card, and proceeds as before. The objective is to maintain an unbroken string of correct answers so as to get from home base to first base, from first base to second base, from second base to third base, and from third base back to home base to score a run. Once a run is scored, the play goes to the other player.

The player with the most runs after all the cards have been drawn wins.

Writing Numerals

Two practice sheets on writing numerals. The one on page 65 is on writing 0, 1, 2, 3, and 4. And the one on page 66 is on writing 5, 6, 7, 8, and 9.

Ziggy's Home Run

A game for two to four on measuring in inches and centimeters.

Materials:

One set of Ziggy's Home Run cards (pp. 307-310) per group. One Ziggy's Home Run gameboard (p. 306), one inch ruler (p. 302), and one centimeter / millimeter ruler (p. 302) per player.

Directions:

To begin, each player draws a card. The player drawing the card with the longest measurement on it plays first. The play rotates clockwise.

To play, a player draws a card and KEEPS it. If it is a card with instructions on it, the player follows the instructions. If it is a card with a measurement on it, the player constructs a segment on his or her gameboard as long as the measurement and labels the segment with its length. If it is the first segment constructed, it must start from somewhere within the home base on the gameboard. If otherwise, it must start from where the last segment left off. In either case, it may point in any direction so long as it does not pass COMPLETELY through a base on the gameboard. Also, it may intersect another segment. The play then goes to the next player.

The objective is to make a "home run" by getting from home base to first base, from first base to second base, from second base to third base, and from third base to home base. To get to a base, a player must construct a segment that ends somewhere within the base. Thus getting from home base to first base might look like the zig-zag below.

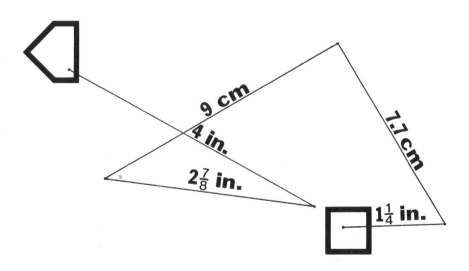

The first player to get to home base wins.

In the event all the Ziggy's Home Run cards are drawn, the cards are collected and shuffled, and the play resumes as before.

Experience Roster

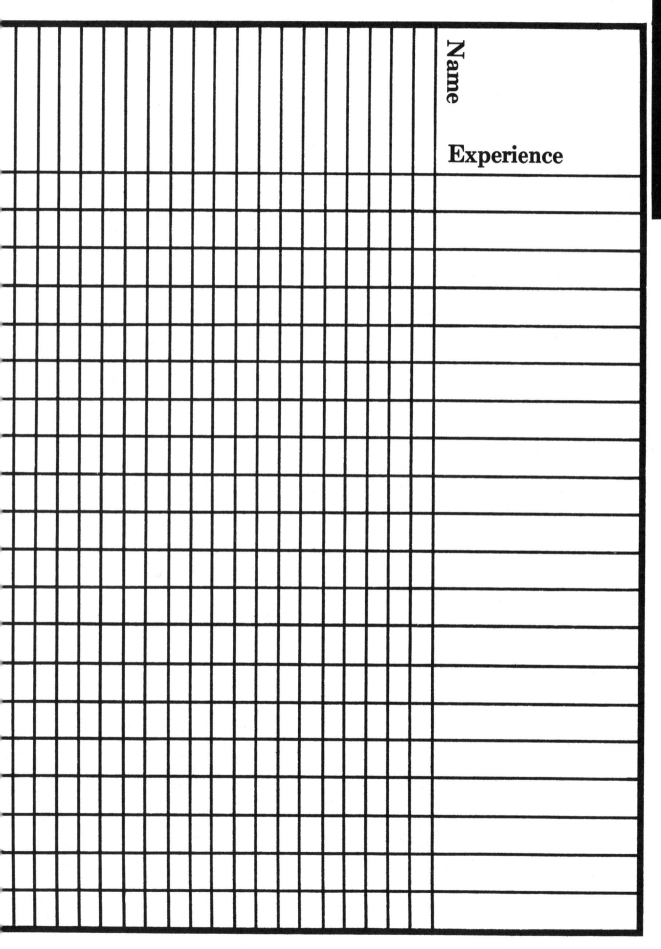

Name	Experience

Spinner Cutouts
(3-, 4-, and 5-sided)
Color, back, and laminate.
Pin to boards or pencils with thumbtacks.

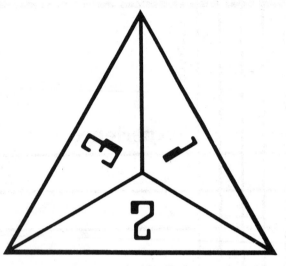

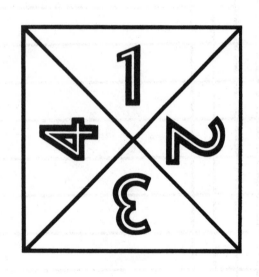

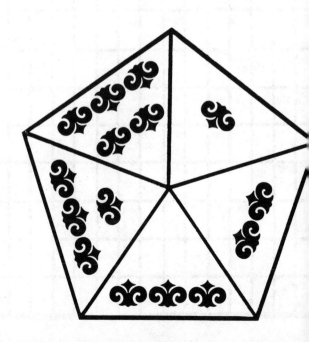

Spinner Cutouts
(6-, 8-, and 10-sided)
Color, back, and laminate.
Pin to boards or pencils with thumbtacks.

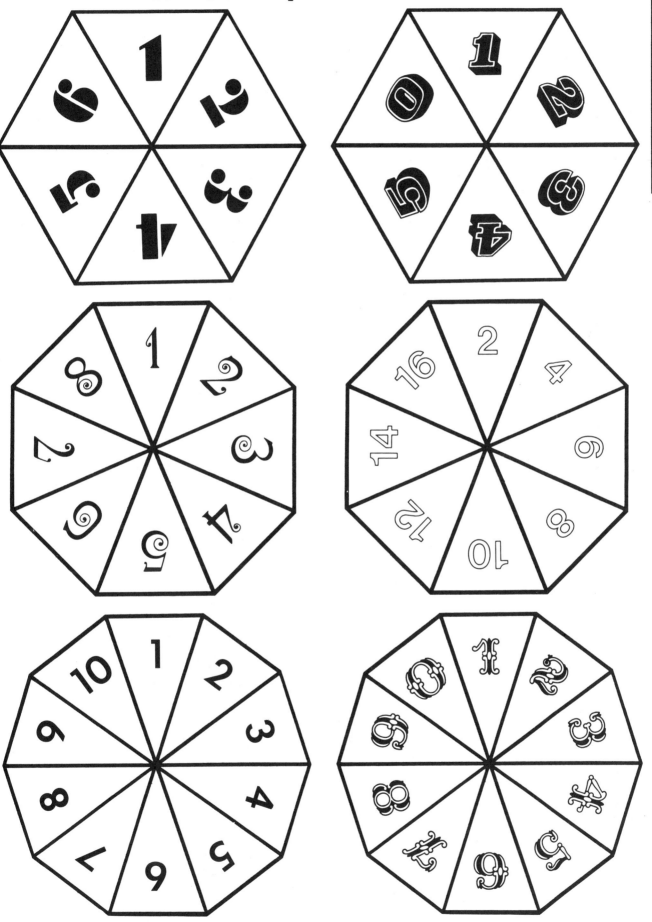

Spinner Cutouts
(Blank)
3-, 4-, and 5-sided

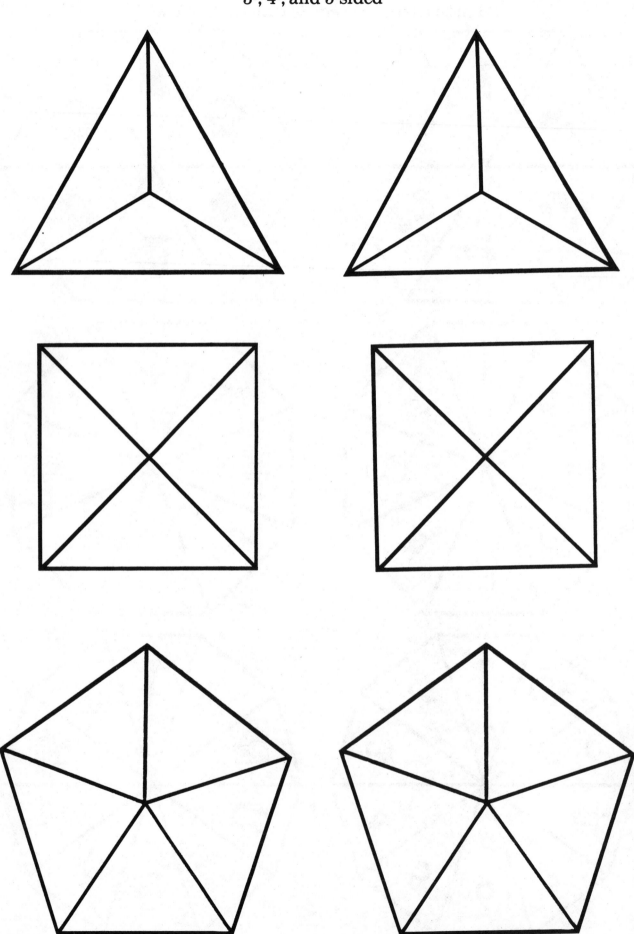

MATH GAMES & ACTIVITIES. COPYRIGHT © 1984

Spinner Cutouts
(Blank)
6-, 8-, and 10-sided

Blank Cards
(6 cm by 9 cm)

Blank Cards
(5 cm by 6 cm)

Blank Cards
(4.5 cm by 4.5 cm)

Blank Tags

Blank Dominoes

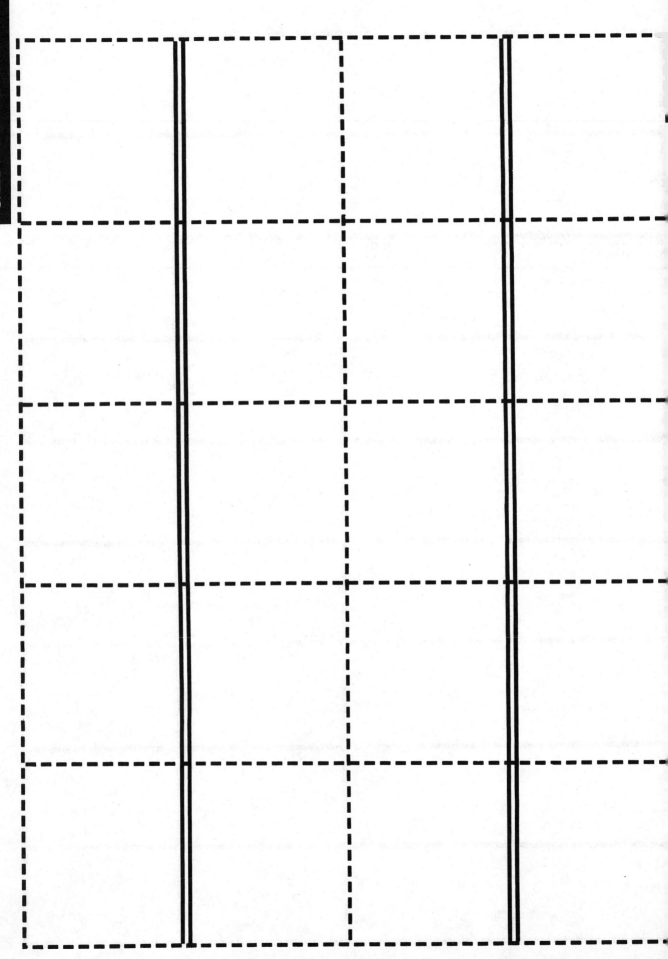

People Pictures
Back, laminate, and cut along solid lines.

Going to the Park Gameboard
(People Pictures)
Color, back, and laminate.

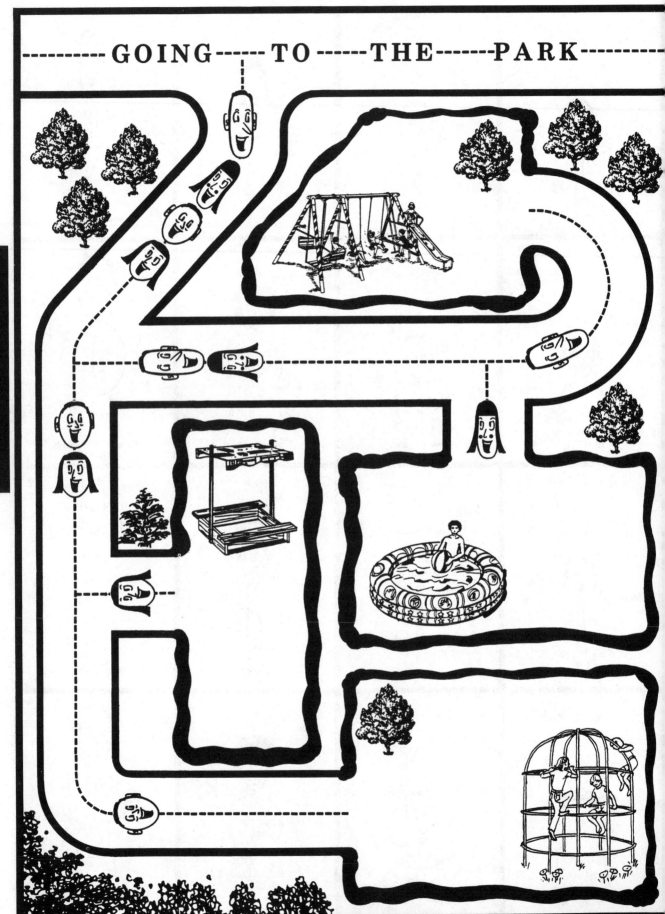

GOING — TO — THE — PARK

CLASSIFICATION

Going to the Park Gameboard
(Blank)

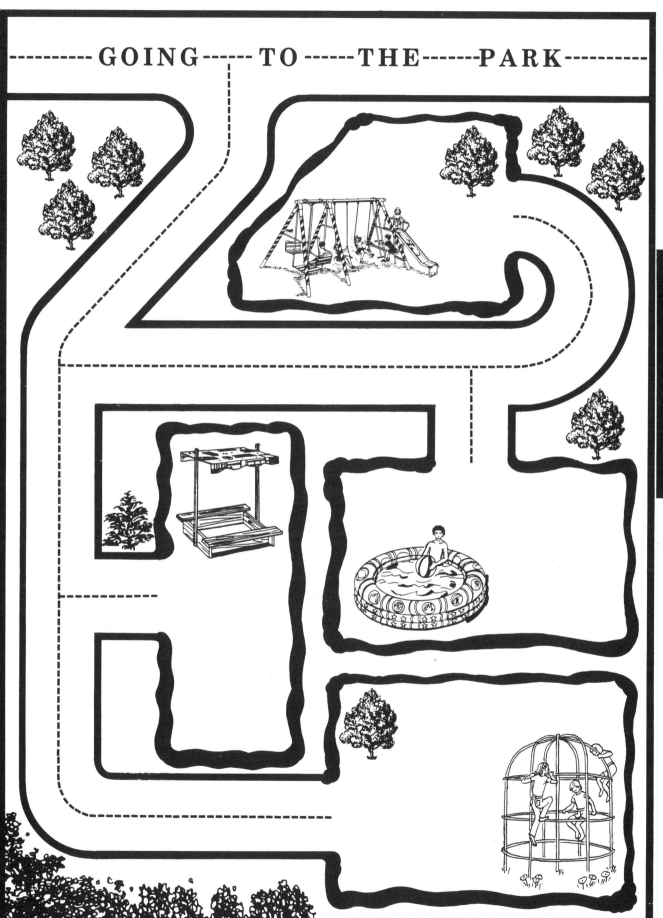

GOING ---- TO ---- THE ---- PARK

Attribute Shape Cutouts
(Page 1 of two pages)
Color, back, and laminate.

CLASSIFICATION

Red

Blue

Yellow

Red

Blue

Yellow

Red

Blue

Yellow

Red

Blue

Yellow

Attribute Shape Cutouts
(Page 2 of two pages)
Color, back, and laminate.

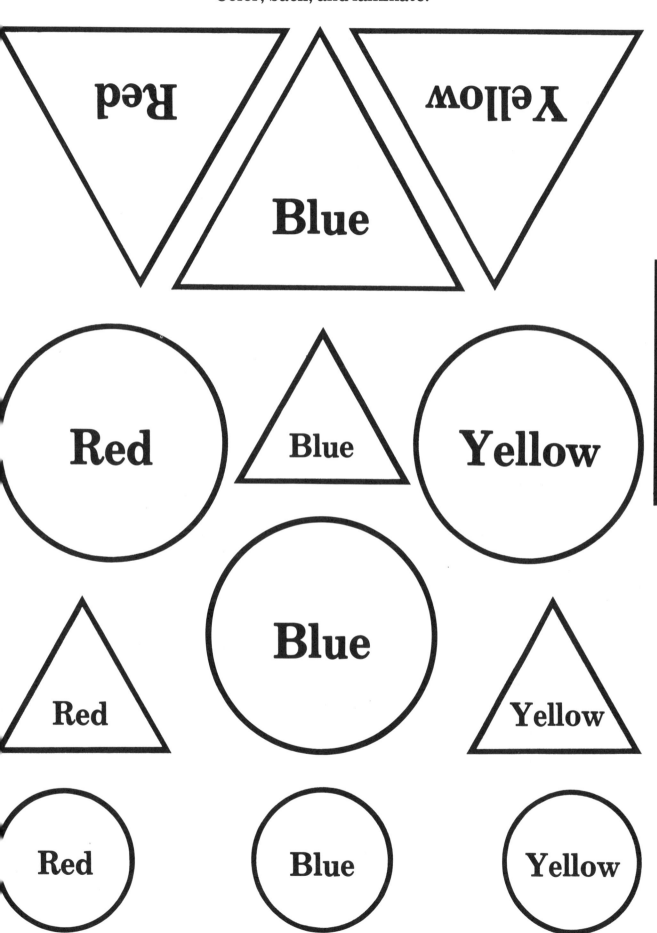

CLASSIFICATION

Gimmi Cards
(Attribute blocks)
Color code the "color" cards.
Back, laminate, and cut along solid lines.

CLASSIFICATION

Big	**Big**	Little	Little
Red	Red	Blue	Blue
Blue	Yellow	Yellow	Yellow
Square	Square	Triangle	Triangle
Circle	Circle	Rectangle	Rectangle

Sameness Trains
(Attribute blocks)

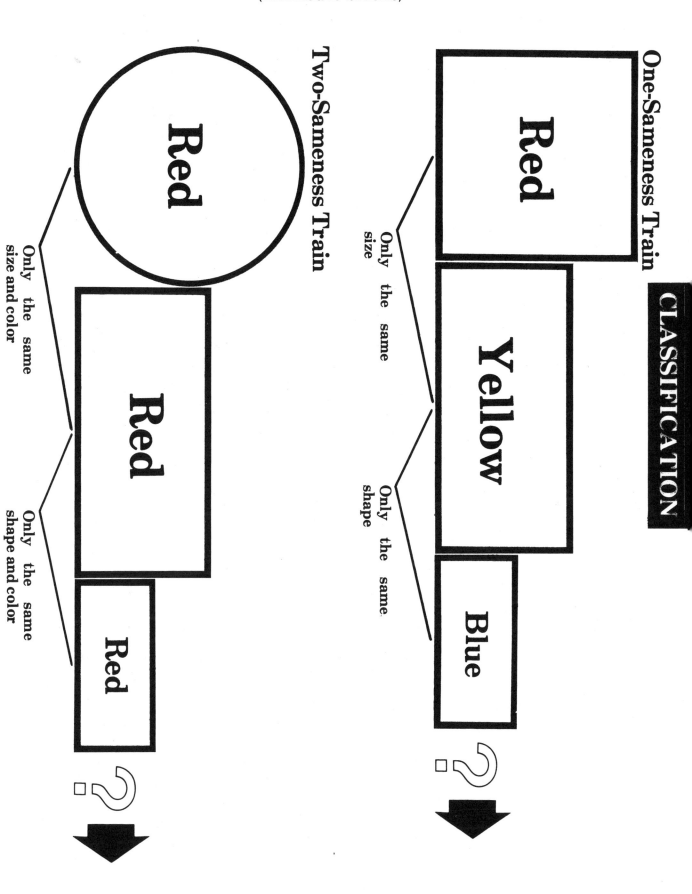

CLASSIFICATION

One-Sameness Train

Red — Only the same size — Yellow — Only the same shape — Blue — ?

Two-Sameness Train

Red — Only the same size and color — Red — Only the same shape and color — Red — ?

Word Cards
(Page 1 of three pages)
Back, laminate, and cut along solid lines.

CLASSIFICATION

I	We	Was
They	Yours	Him
The	For	Had
Any	Now	His
And	Your	My

Word Cards
(Page 2 of three pages)
Back, laminate, and cut along solid lines.

She	Here	Got
To	Are	Our
Has	Them	Say
A	Not	On
Us	See	Two

CLASSIFICATION

Word Cards
(Page 3 of three pages)
Back, laminate, and cut along solid lines.

CLASSIFICATION

You	But	He
Day	How	Have
Of	All	Get
Out	Did	In
Me	Can	Her

One-letter word	Begins with A	Begins with G	Begins with H
Two-letter word	Begins with I	Begins with O	Begins with T
Three-letter word	Begins with Y	Ends in D	Ends in E
Four-letter word	Ends in N	Ends in R	Ends in S
Five-letter word	Ends in T	Ends in W	Ends in Y

CLASSIFICATION

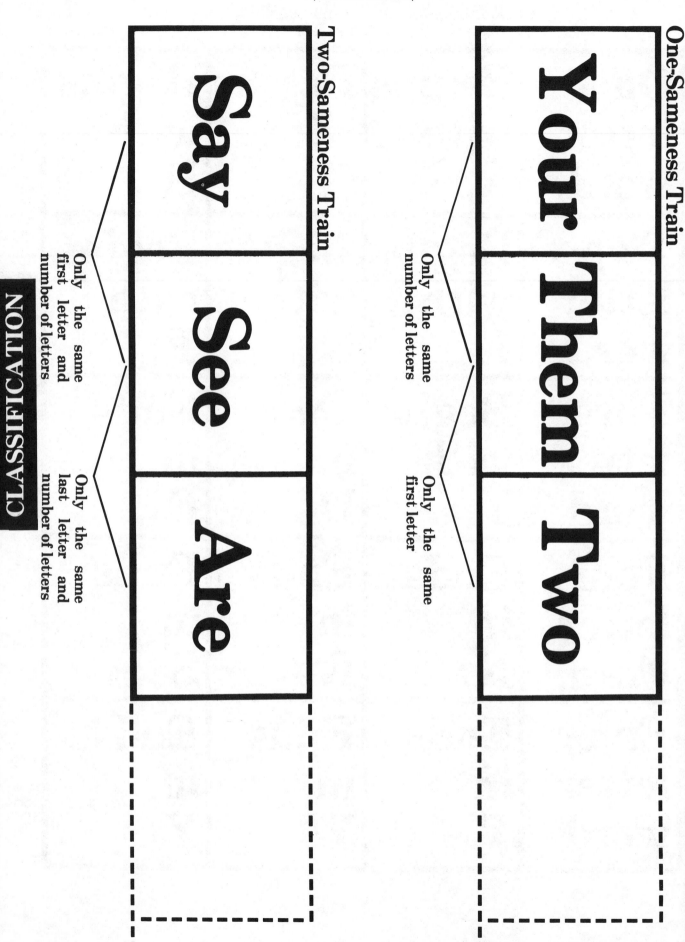

One-Sameness Train

Your

Them

Only the same
number of letters

Two

Only the same
first letter

Two-Sameness Train

Say

See

Only the same
first letter and
number of letters

Are

Only the same
last letter and
number of letters

CLASSIFICATION

Two-Way Sort Task Card
Label, back, and laminate.

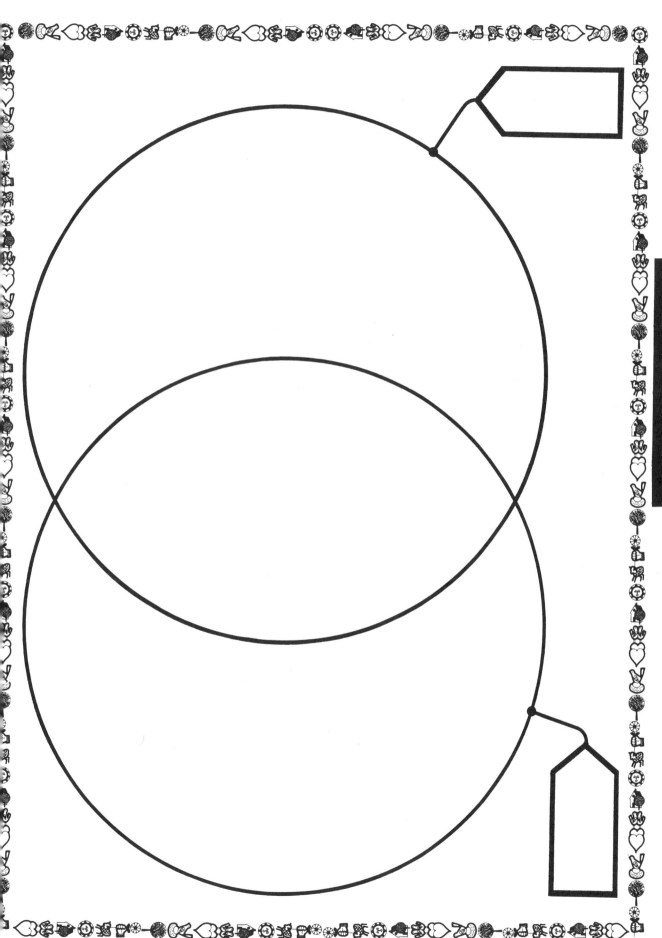

CLASSIFICATION

CLASSIFICATION

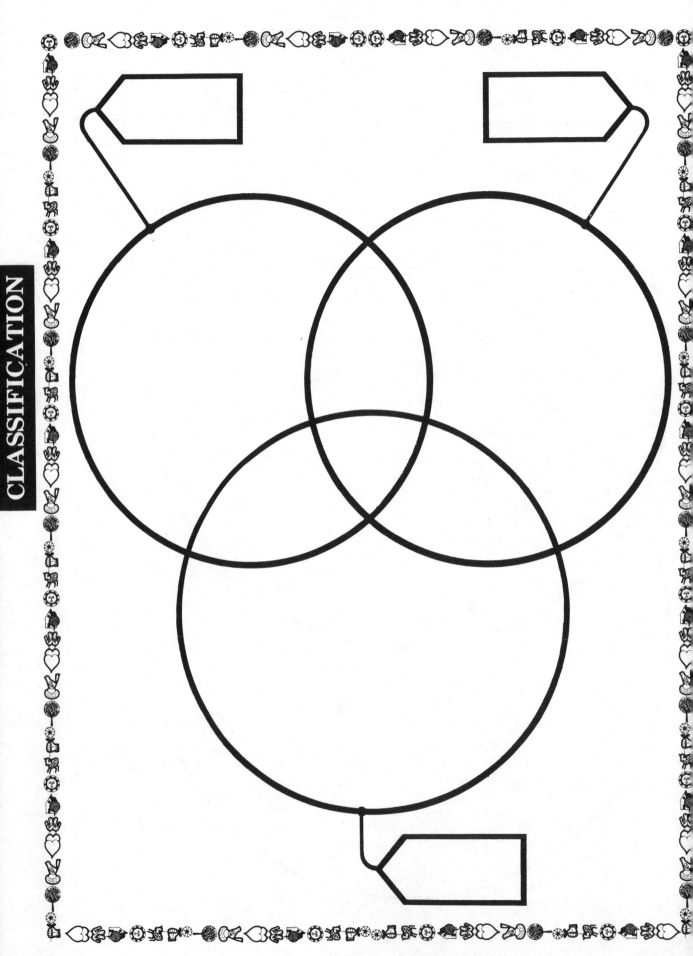

Car Park Gameboard
Color, back, and laminate.

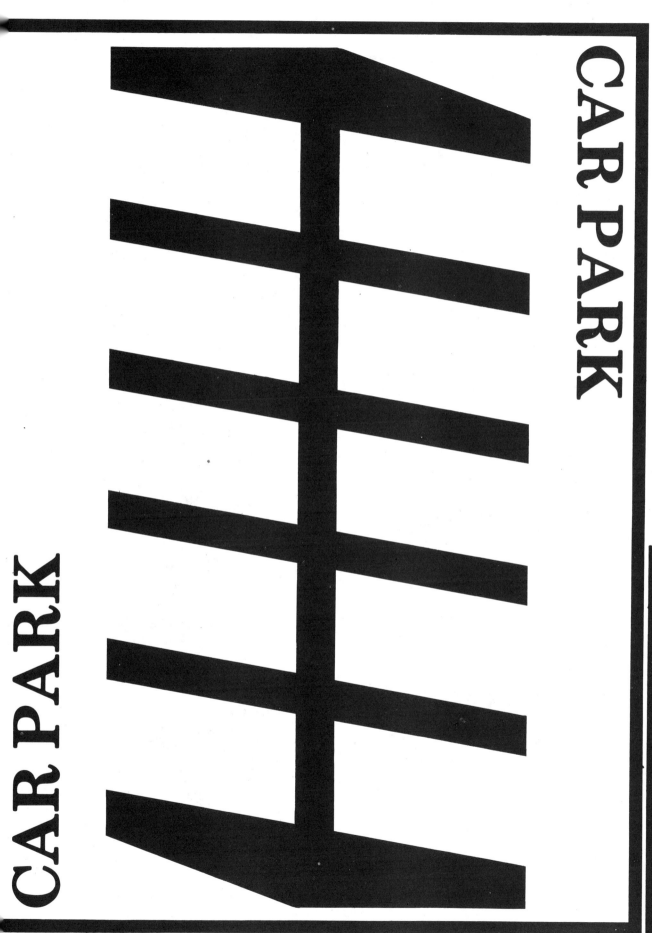

CAR PARK

CAR PARK

Car Park Cutouts
Color, back, and laminate.

Watermelon Gameboard
Color, back, and laminate.

Pick-a-Pair Cards
(Page 1 of three pages)
Color, back, and laminate.
Cut along solid lines.

ONE-TO-ONE CORRESPONDENCE

Pick-a-Pair Cards
(Page 2 of three pages)
Color, back, and laminate.
Cut along solid lines.

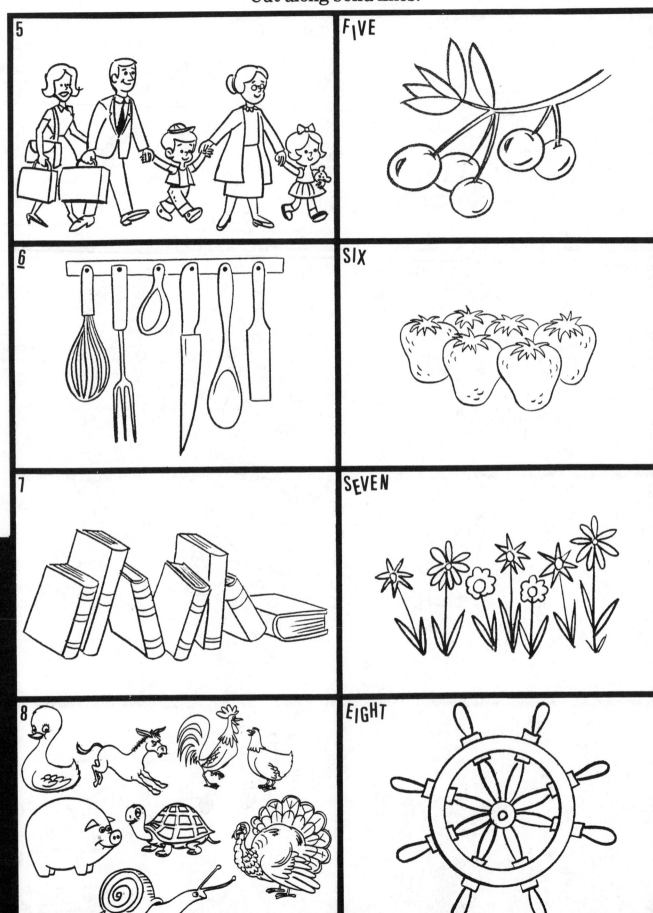

5

FIVE

6

SIX

7

SEVEN

8

EIGHT

ONE-TO-ONE CORRESPONDENCE

Pick-a-Pair Cards
(Page 3 of three pages)
Color, back, and laminate.
Cut along solid lines.

Double Nine Dominoes
(Page 1 of six pages)
Back, laminate, and cut along dotted lines.

ONE-TO-ONE CORRESPONDENCE

Double Nine Dominoes
(Page 2 of six pages)
Back, laminate, and cut along dotted lines.

ONE-TO-ONE CORRESPONDENCE

Double Nine Dominoes
(Page 3 of six pages)
Back, laminate, and cut along dotted lines.

ONE-TO-ONE CORRESPONDENCE

34

Double Nine Dominoes
(Page 4 of six pages)
Back, laminate, and cut along dotted lines.

Double Nine Dominoes
(Page 5 of six pages)
Back, laminate, and cut along dotted lines.

ONE-TO-ONE CORRESPONDENCE

Double Nine Dominoes
(Page 6 of six pages)
Back, laminate, and cut along dotted lines.

ONE-TO-ONE CORRESPONDENCE

Block It or Shade In Gameboards
(6 by 6)
Back and Laminate.
(Consumable for Shade In)

Block It or Shade In Gameboard
(12 by 12)
Back and laminate.
(Consumable for Shade In)

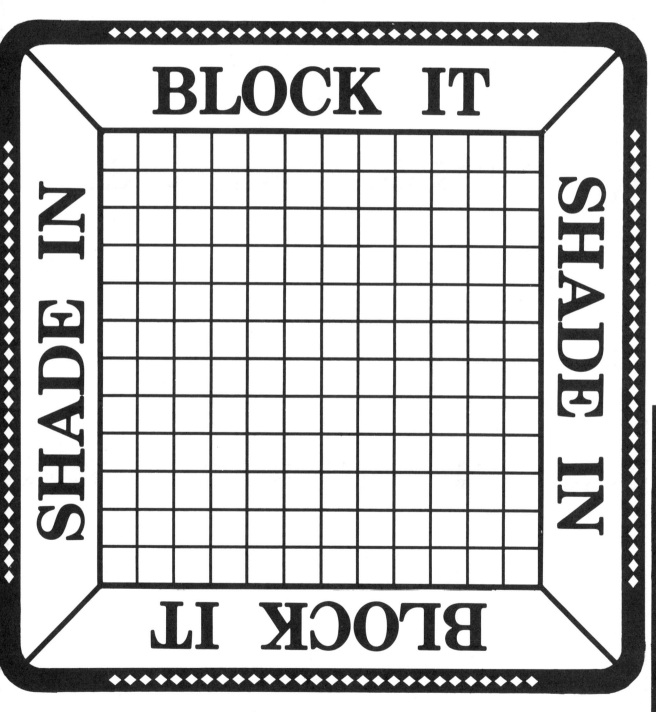

ONE-TO-ONE CORRESPONDENCE

Sequence Cards
(Pencil)
Color, back, and laminate.
Cut along solid lines.

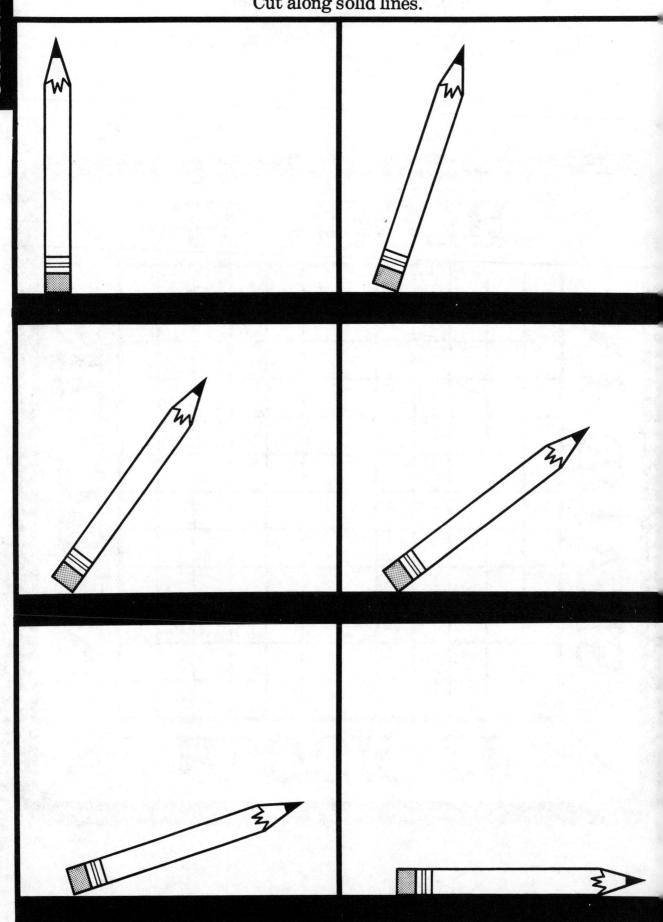

Sequence Cards
(Boat)
Color, back, and laminate.
Cut along solid lines.

Sequence Cards
(Ball)
Color, back, and laminate.
Cut along solid lines.

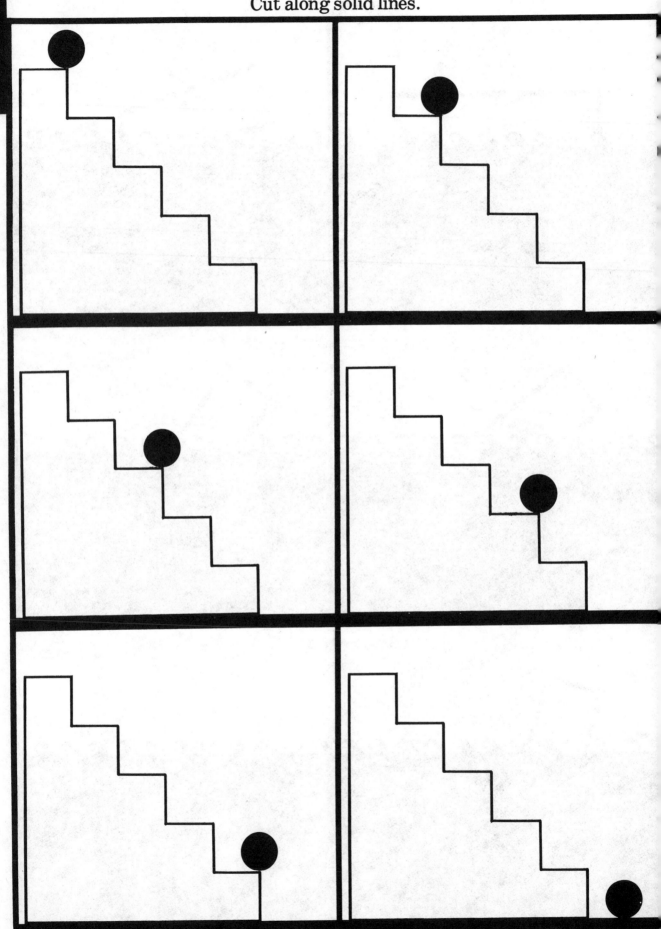

MATH GAMES & ACTIVITIES. COPYRIGHT © 1984

Sequence Cards
(Liquid)
Color, back, and laminate.
Cut along solid lines.

Rod Patterns
(Steps)

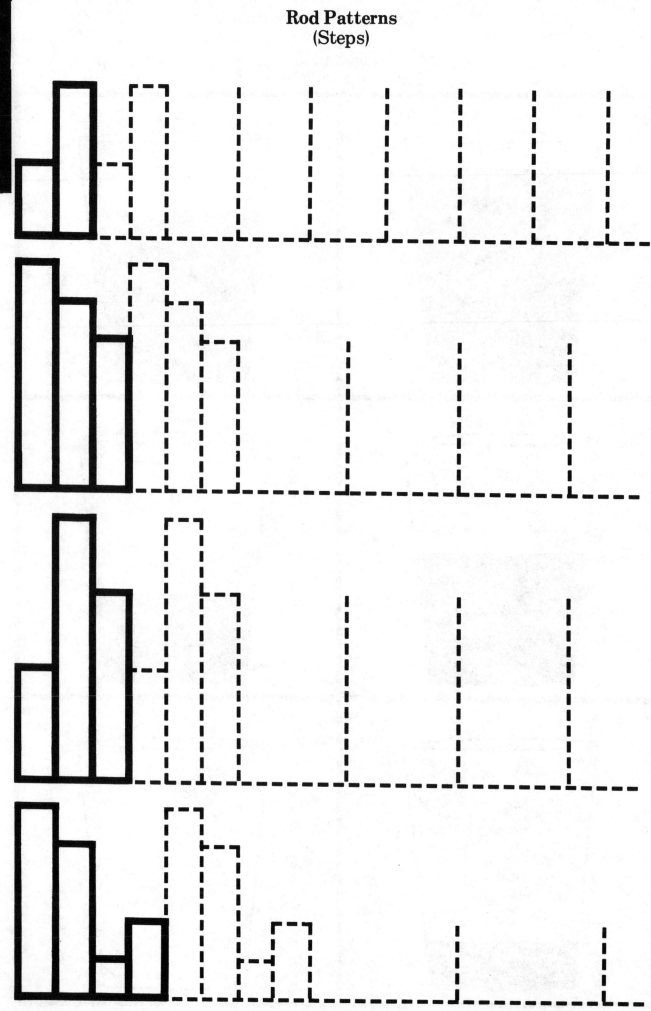

Rod Patterns
(Lines)

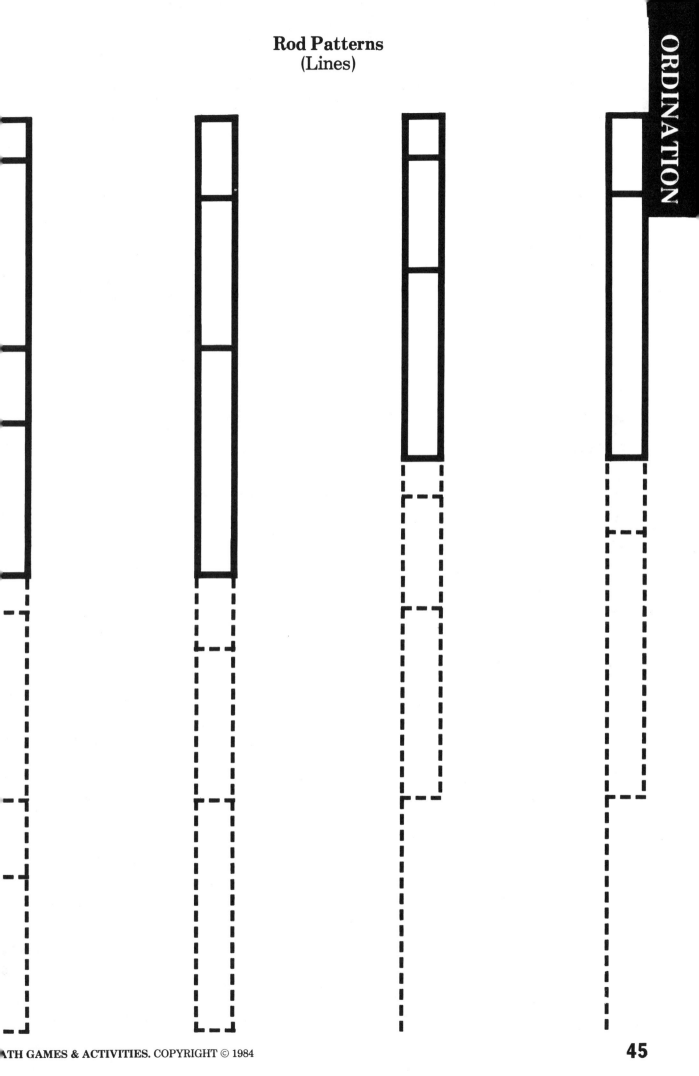

Bridge Gameboard
Color, back, and laminate.

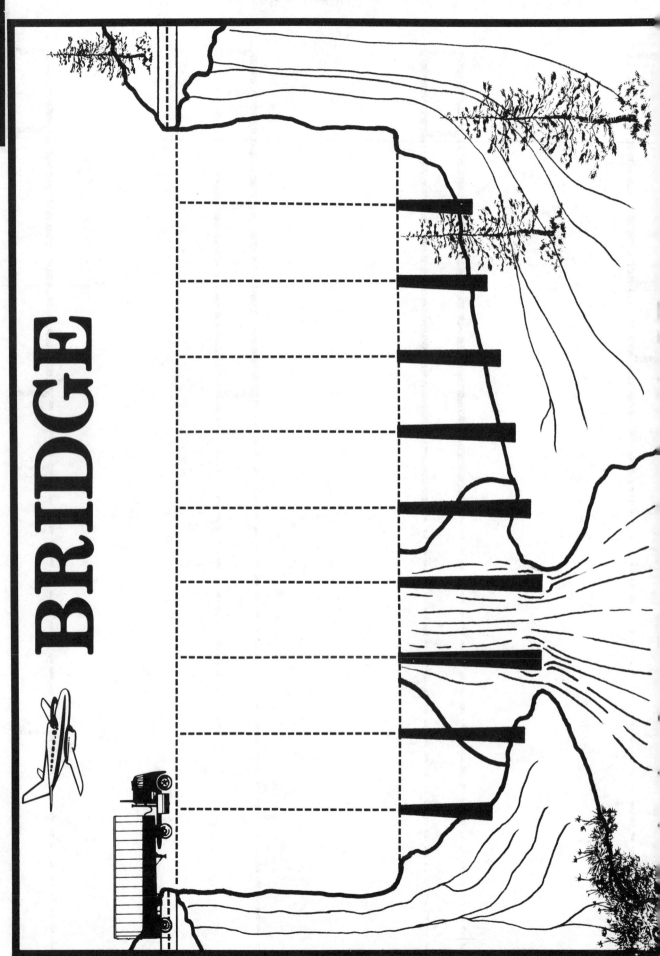

BRIDGE

Bridge Cutouts
Color, back, and laminate.

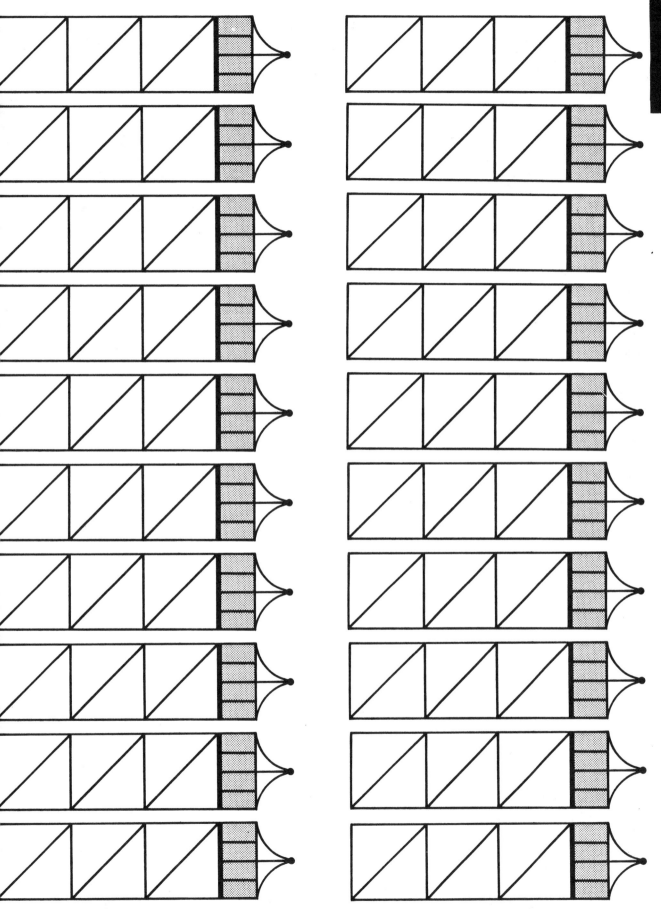

Lily Pad Gameboard
Color, back, and laminate

LILY PAD

Lily Pad Cutouts
Color, back, and laminate.

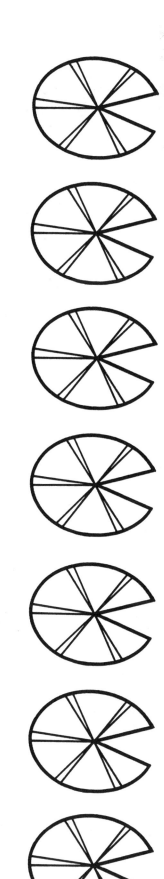

Hex Gameboard
Color, back, and laminate.

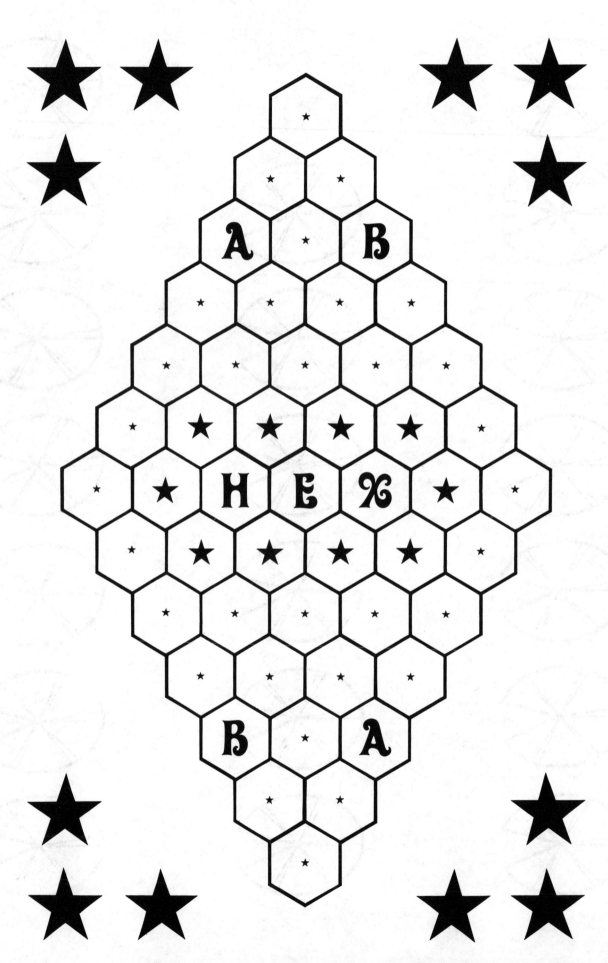

  MATH GAMES & ACTIVITIES. COPYRIGHT © 1984

Hi!   My name is Ima Blank-
page.   Turn   the   page   and
BEHOLD!

COUNTING

BALANCING ACT

52

Balancing Act Gameboard
Right side
Color, back, and laminate.

COUNTING

Clown Gameboard
Right side
Color, back, and laminate.

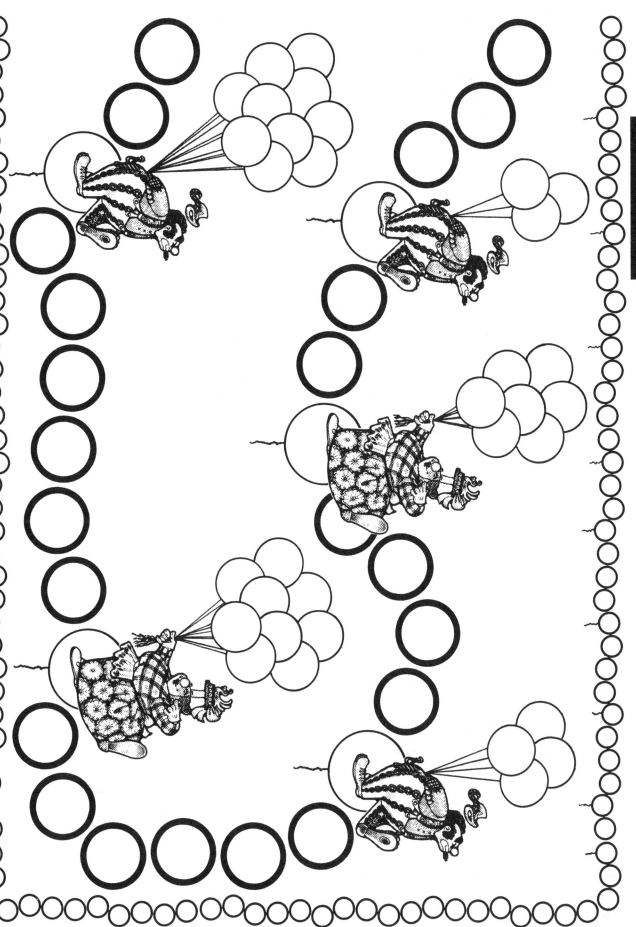

COUNTING

Safari Gameboard
Right side
Color, back, and laminate.

Safari Cutouts
Color, back, and laminate.

58   **MATH GAMES & ACTIVITIES.** COPYRIGHT © 1984

Collect-a-Shape Gameboard
Color, back, and laminate.

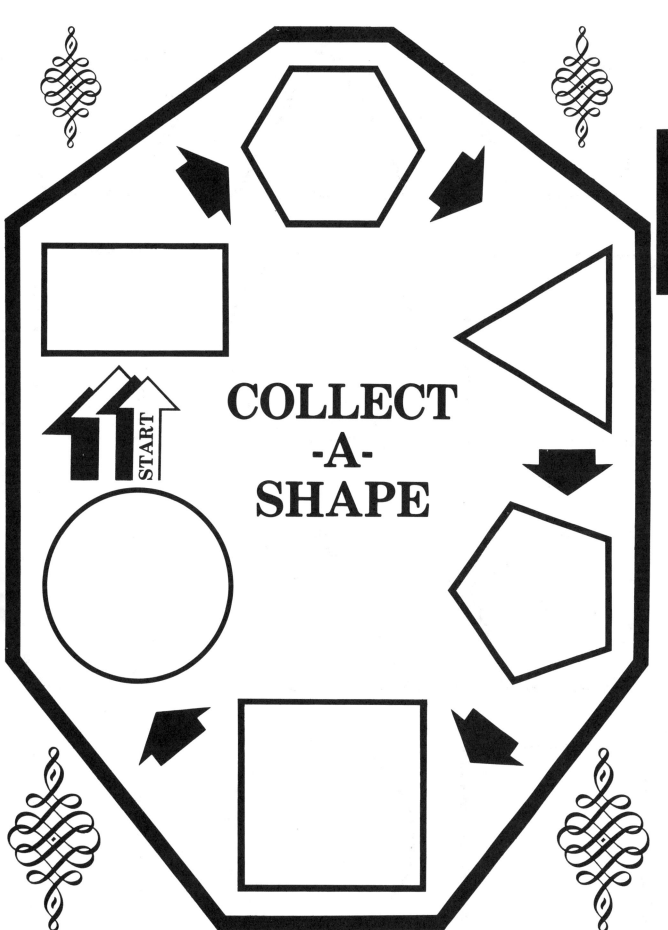

COLLECT
-A-
SHAPE

START

Collect-a-Shape Cutouts
Color, back, and laminate.

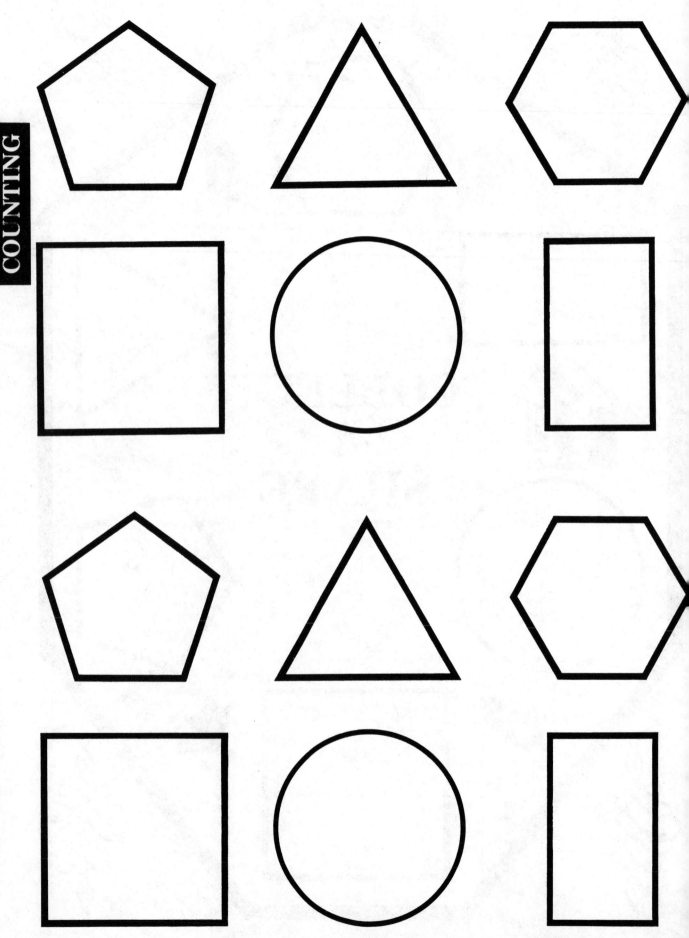

MATH GAMES & ACTIVITIES. COPYRIGHT © 1984

Hi! My name is Ima Blank-page. Turn the page and BEHOLD!

Telephone Gameboard
Left side
Color, back, and laminate.

448-1795

629-3284

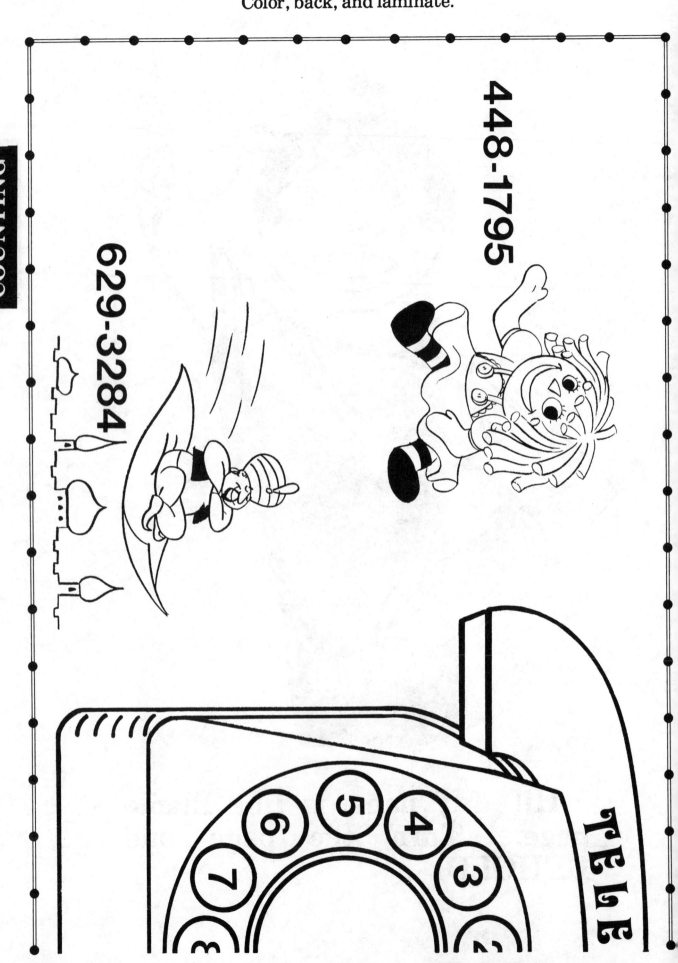

Telephone Gameboard
Right side
Color, back, and laminate.

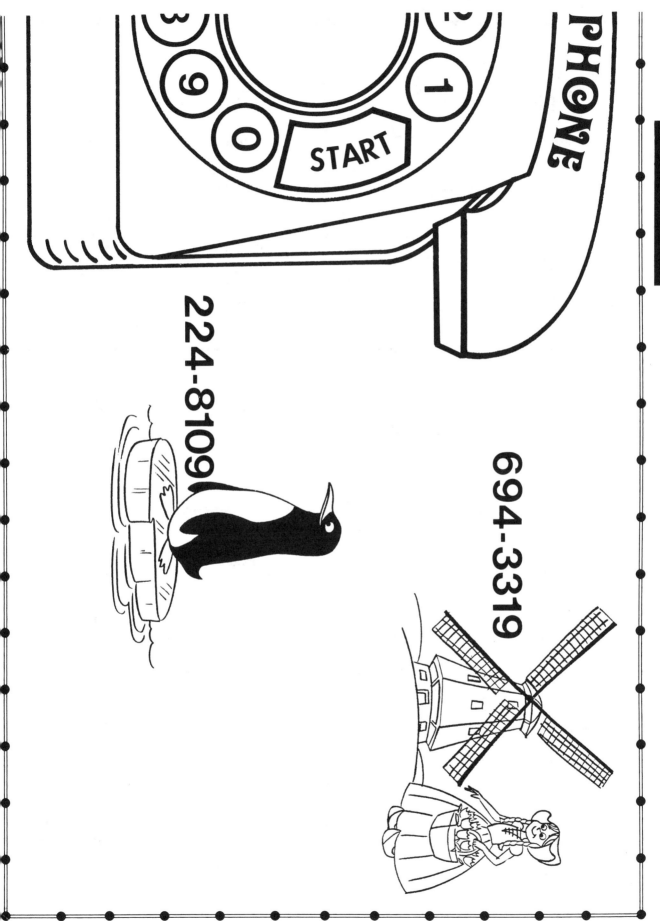

START

PHONE

224-8109

694-3319

Telephone Numerals
Back, laminate, and cut along solid lines.

0	0	1	1
2	2	3	3
4	4	5	5
6	6	7	7
8	8	9	9

Name . . . .

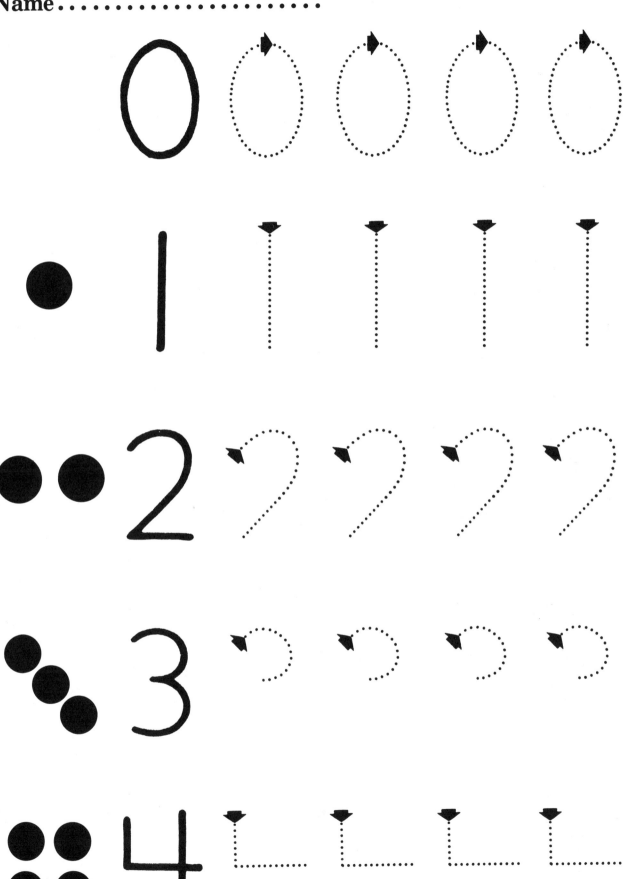

NUMBER NAMES

Name......................

NUMBER NAMES

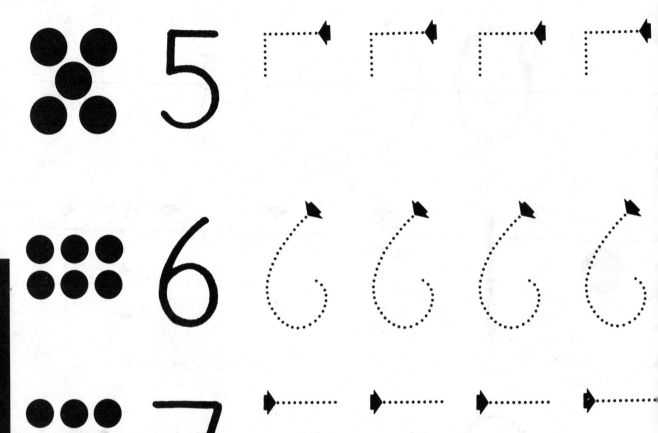

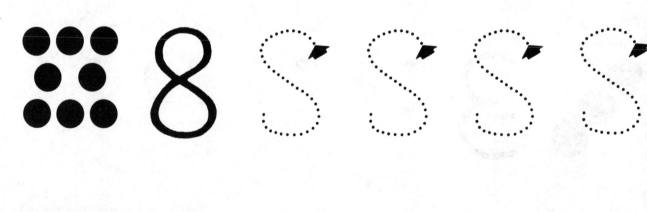

Numeral Puzzles
(1-4)
Back, laminate, and cut along solid lines.

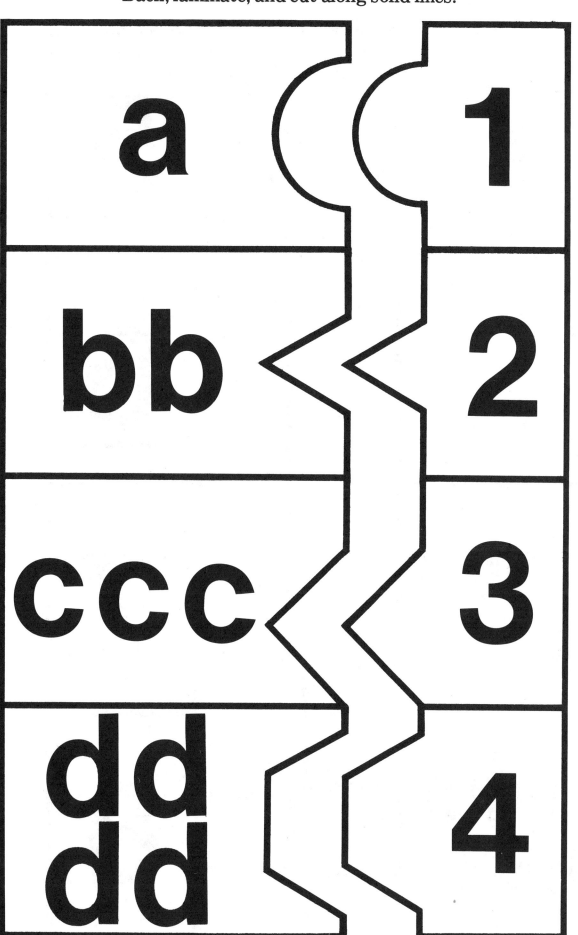

a

b b

c c c

d d
d d

1

2

3

4

NUMBER NAMES

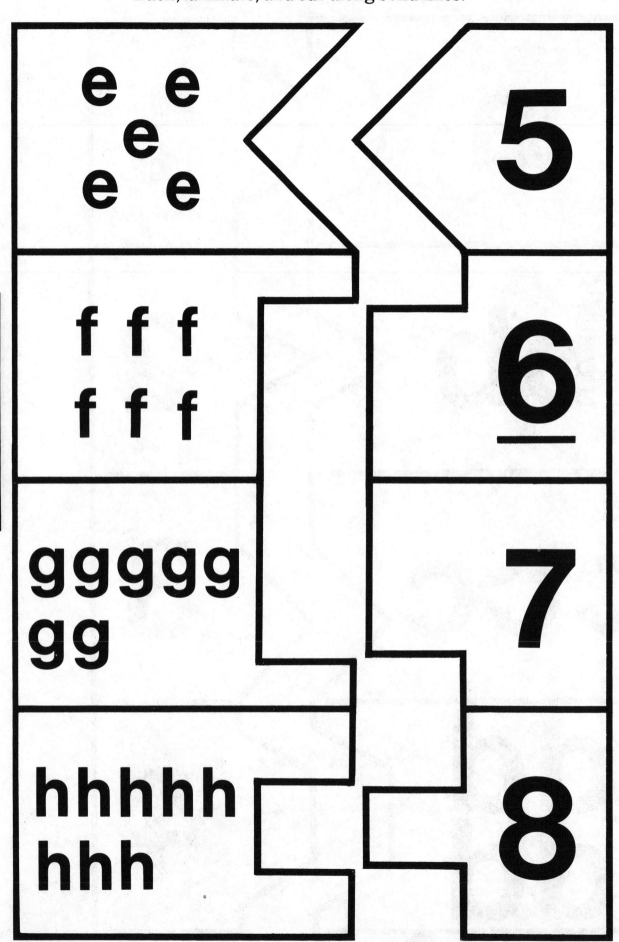

Numeral Puzzles
(9-12)
Back, laminate, and cut along solid lines.

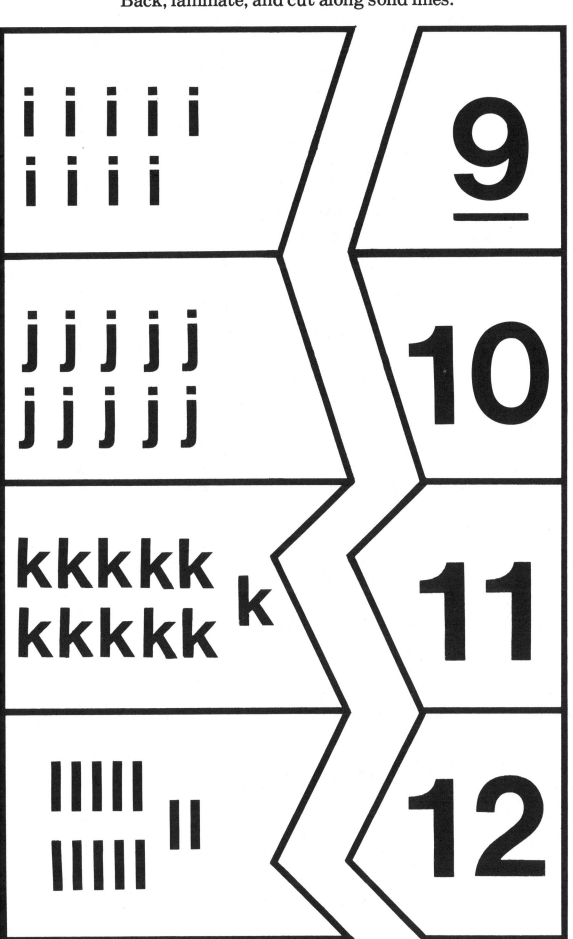

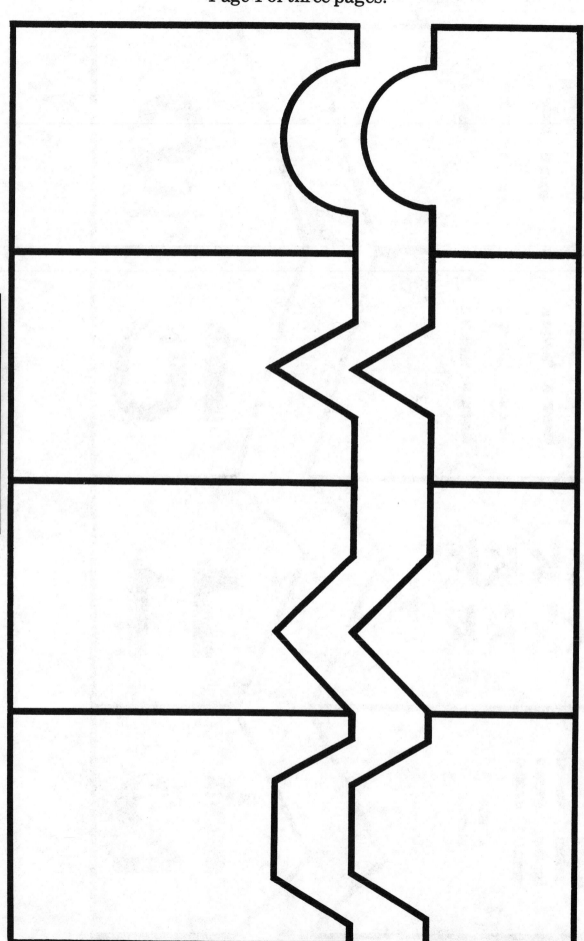

NUMBER NAMES

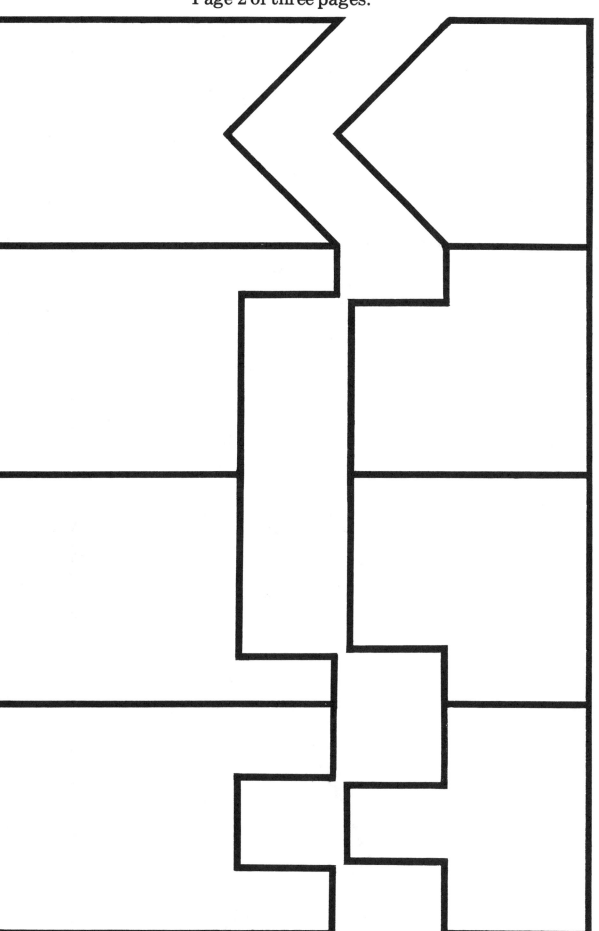

NUMBER NAMES

NUMBER NAMES

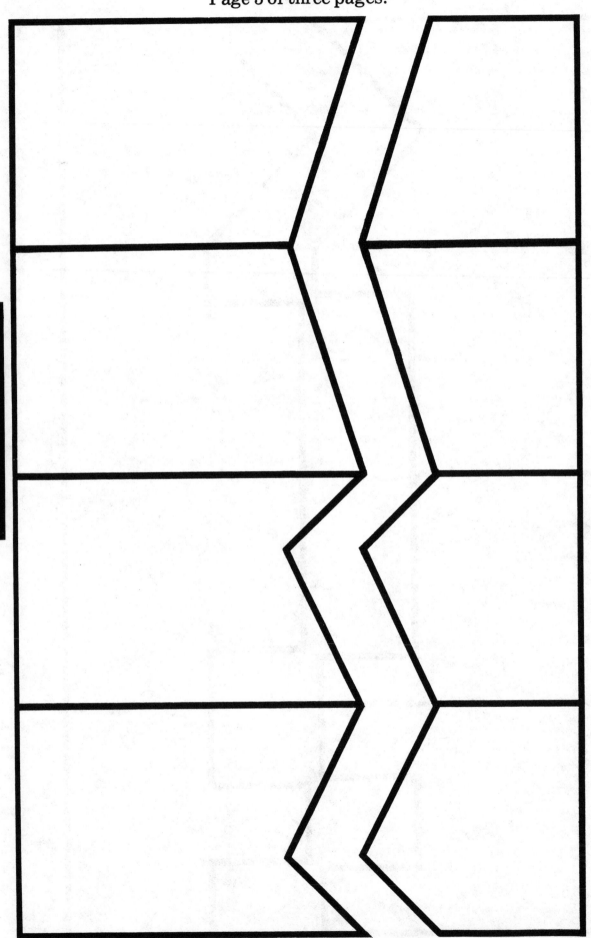

Name..............................

Match the smoke rings with the pipes.

NUMBER NAMES

Name................................

Match the bubbles with the fish.

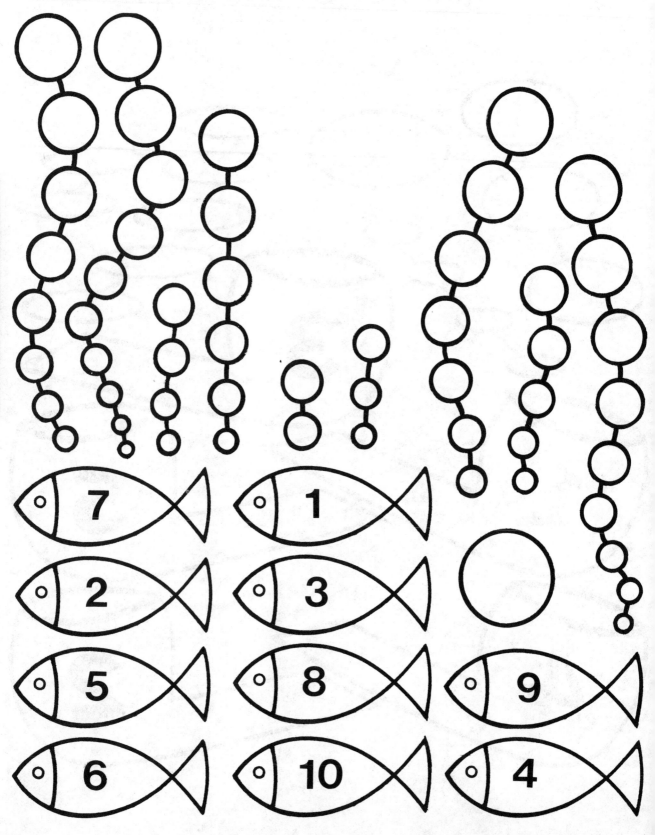

NUMBER NAMES

Numeral Cards
(1-3 / one-three)
Back, laminate, and cut along solid lines.

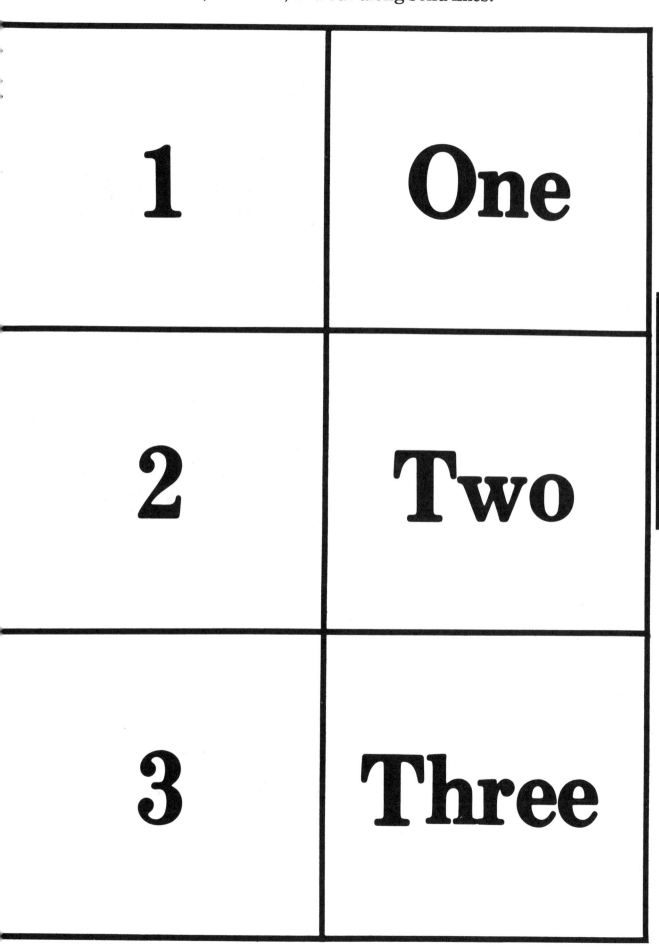

1	One
2	Two
3	Three

NUMBER NAMES

4	**Four**
5	**Five**
6	**Six**

Numeral Cards
(7-9 / seven-nine)
Back, laminate, and cut along solid lines.

7	**Seven**
8	**Eight**
<u>9</u>	**Nine**

77

NUMBER NAMES

10	Ten
11	Eleven
12	Twelve

Action Cards
(Page 1 of two pages)
Back, laminate, and cut along solid lines.

Clap	Hop
Wink	Nod
Shake	Point
Tap	Wave

NUMBER NAMES

Action Cards
(Page 2 of two pages)
Back, laminate, and cut along solid lines.

NUMBER NAMES

Laugh	Squirm
Smile	Touch
Feel	Grin
Blink	Wiggle

Numeral Dominoes
(Page 1 of six pages)
Back, laminate, and cut along dotted lines.

Zero	0	0	1
0	3	0	4
0	6	0	7
0	9	1	One
1	3	1	4

NUMBER NAMES

Numeral Dominoes
(Page 2 of six pages)
Back, laminate, and cut along dotted lines.

NUMBER NAMES

3	6	3	7
3	6	Four	4
4	6	4	7
4	9	5	Five
5	7	5	8

Numeral Dominoes
(Page 3 of six pages)
Back, laminate, and cut along dotted lines.

4	3	3	Three
8	2	7	2
5	2	4	2
Two	2	$\frac{9}{}$	1
Eight	8	$\frac{9}{}$	7

NUMBER NAMES

Numeral Dominoes
(Page 4 of six pages)
Back, laminate, and cut along dotted lines.

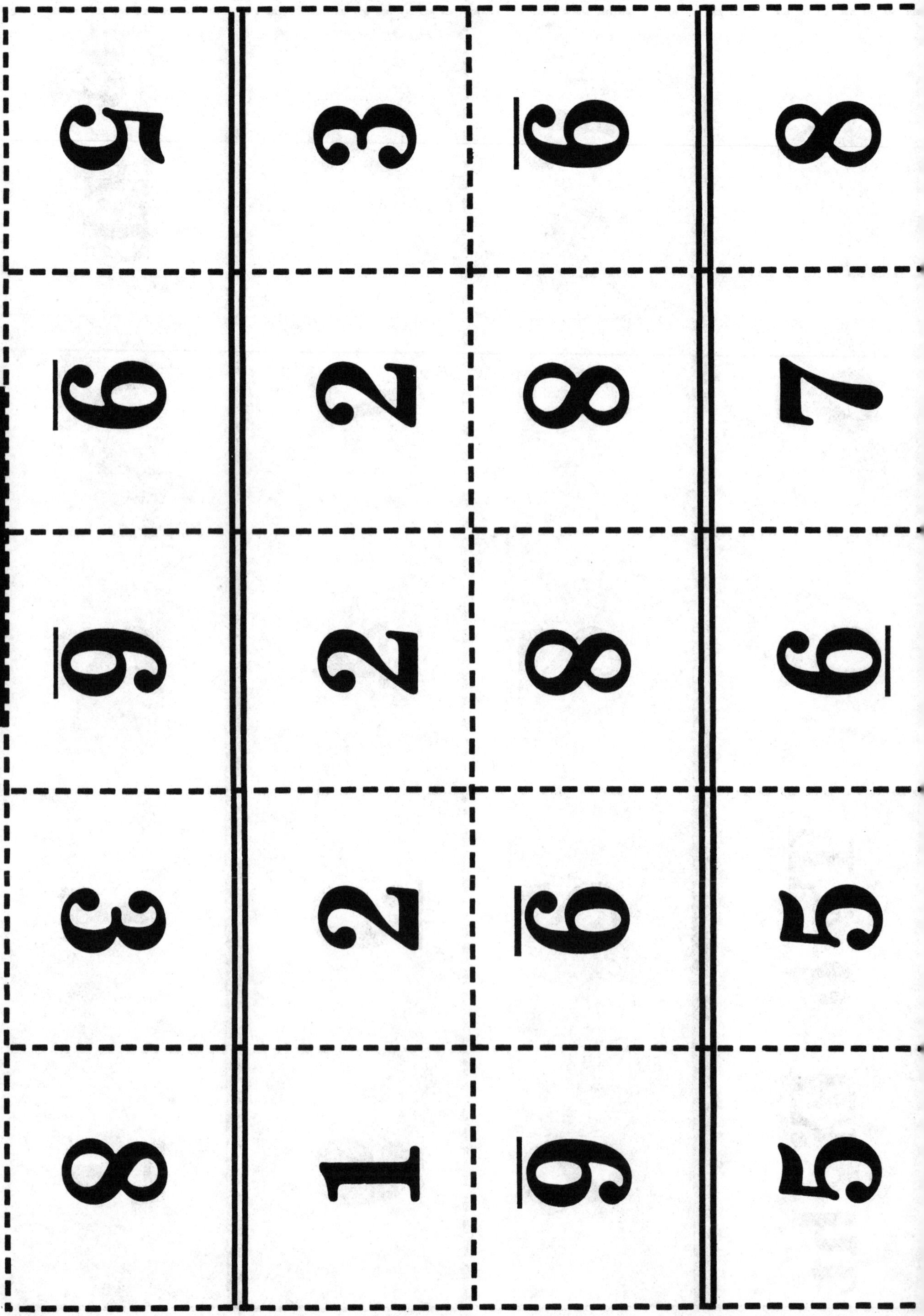

NUMBER NAMES

5	3	6̅	8
9̲	2	8	7
6̅	2	8	6̲
3	2	6̅	5
8	1	9̅	5

84

Numeral Dominoes
(Page 5 of six pages)
Back, laminate, and cut along dotted lines.

NUMBER NAMES

2	1	3	8
8	0	4	5
5	0	9	Nine
1	6	1	7
Six	6	6	7

Numeral Dominoes
(Page 6 of six pages)
Back, laminate, and cut along dotted lines.

NUMBER NAMES

7	Seven	6	6
5	1	8	4
2	0		

Name............................

What am I? To find out, connect the dots in order.

Name. . . . .

What am I? To find out, connect the dots in order.

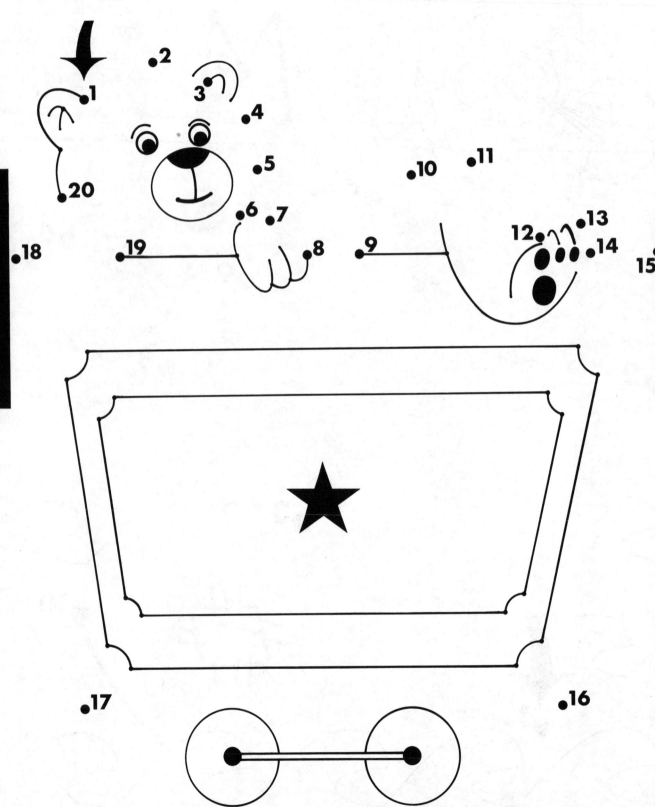

NUMBER NAMES

Connect-a-Dot
(By twos)

Name............................

What am I? Count by twos to find out.

NUMBER NAMES

 8

6 10 4

74

72

70

76

68

66

64

12 26

2 62 58 60 78 80 54

14 56

16

24

22

28 52

18 50

20 48

30 40

42

46

44

38

36

32 34

Name..............................

What am I? Count by fives to find out.

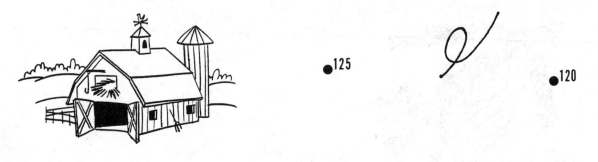

NUMBER NAMES

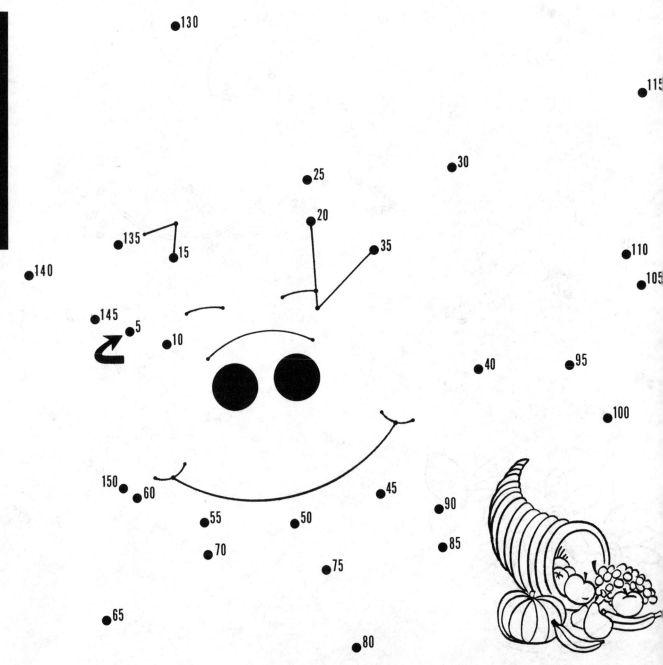

●125

●120

●130

●115

●30

●25

●20

●35

●135

●15

●140

●110

●105

●145

●5

●10

●40

●95

●100

●150 ●60

●45

●55 ●50

●90

●70

●85

●75

●65

●80

90   MATH GAMES & ACTIVITIES. COPYRIGHT © 1984

Name...

One more   One less   Between

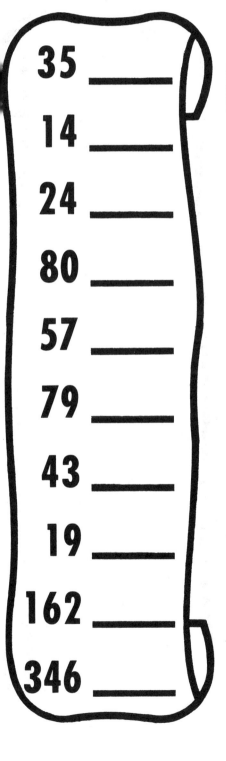
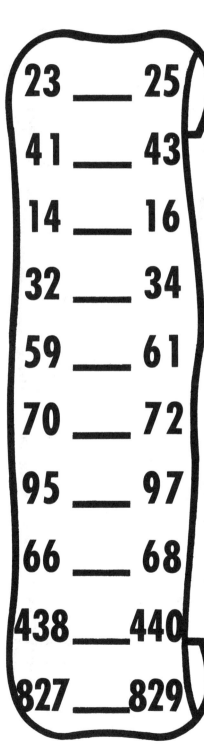

One more	One less	Between
35 ___	___ 34	23 ___ 25
14 ___	___ 28	41 ___ 43
24 ___	___ 42	14 ___ 16
80 ___	___ 57	32 ___ 34
57 ___	___ 11	59 ___ 61
79 ___	___ 18	70 ___ 72
43 ___	___ 49	95 ___ 97
19 ___	___ 73	66 ___ 68
162 ___	___ 260	438 ___ 440
346 ___	___ 586	827 ___ 829

NUMBER NAMES

Name..

One more One less Between

Multi-base Block Cutouts
(Large units and longs)
Color, back, and laminate.

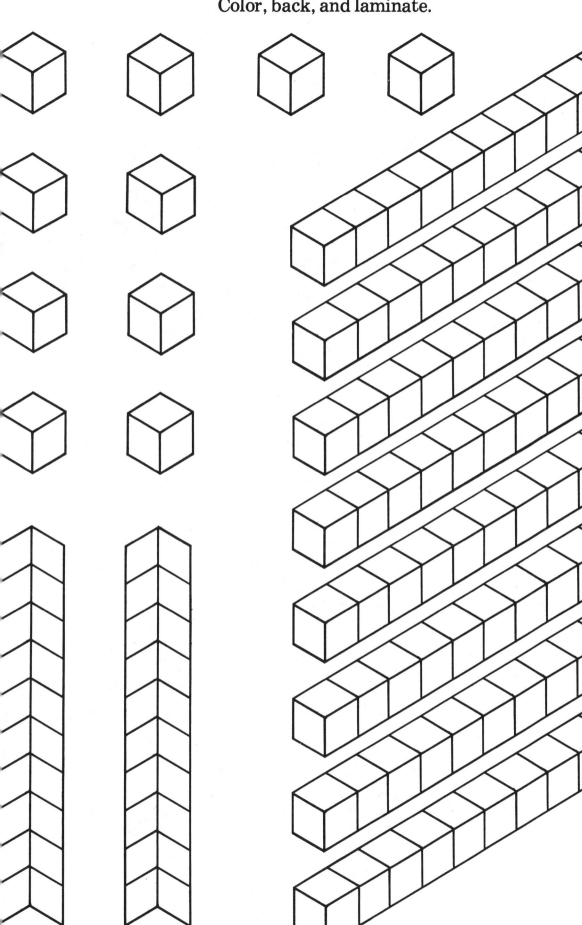

Multi-base Block Cutouts
(Large flats)
Color, back, and laminate.

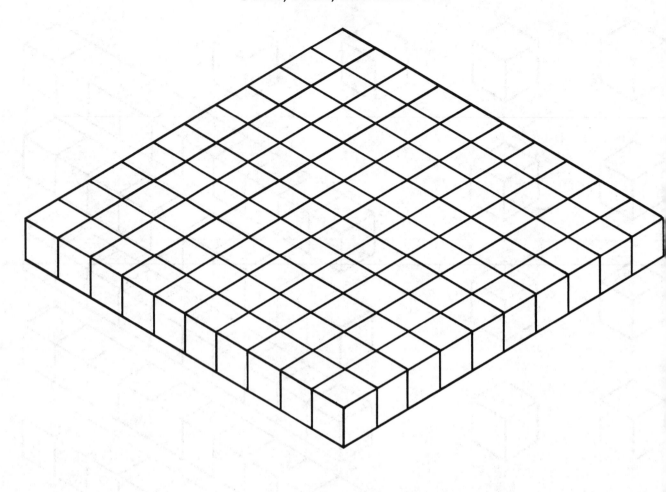

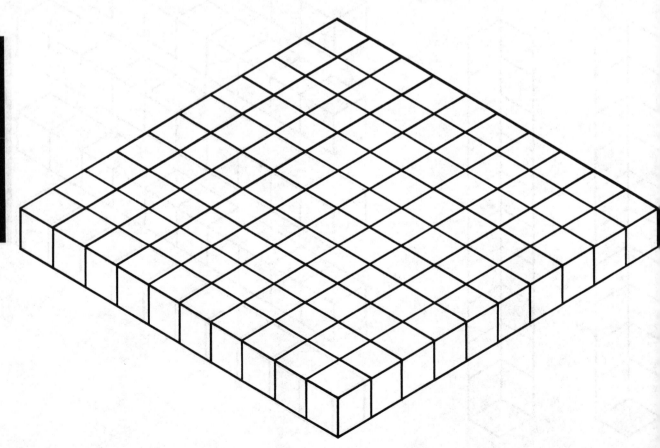

Multi-base Block Cutout
(Large cube)
Color, back, and laminate.

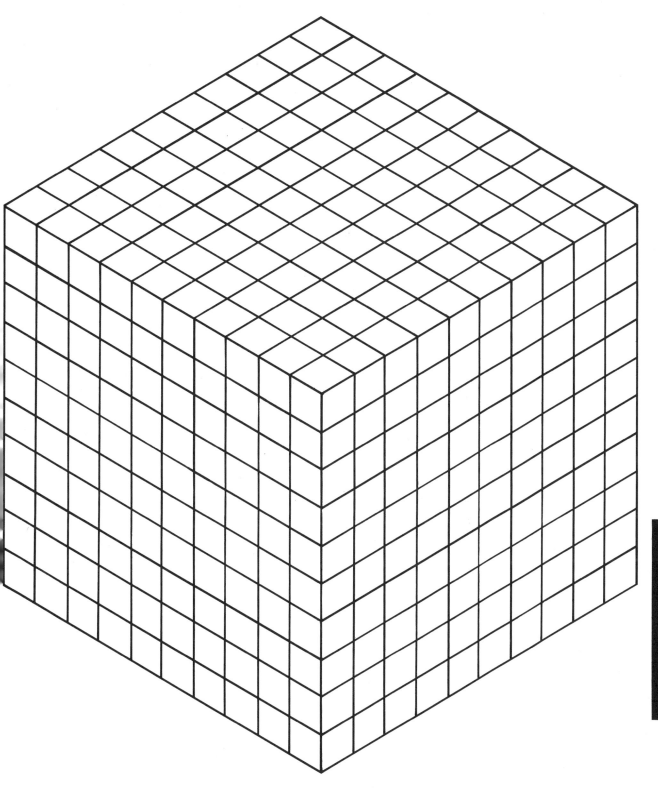

Multi-base Block Cutouts
(Small units and longs)

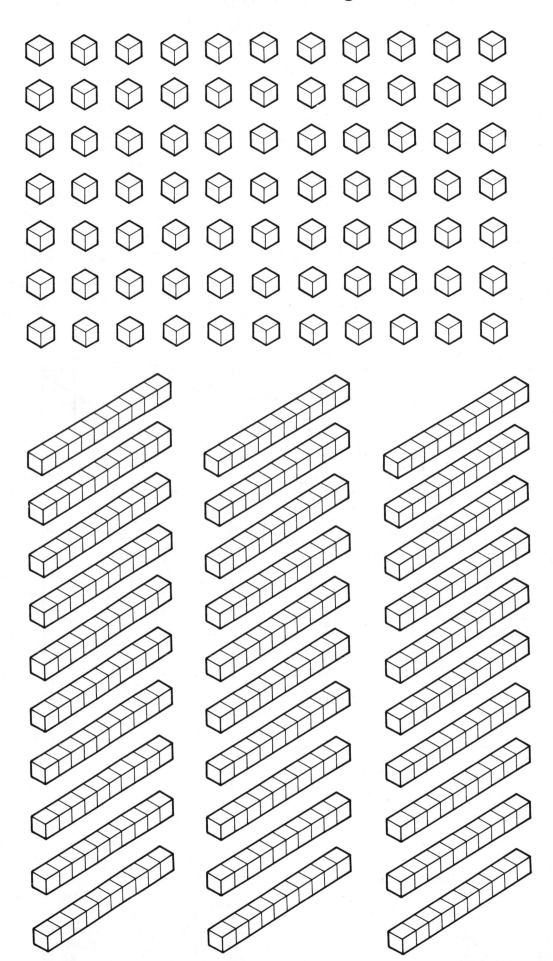

Multi-base Block Cutouts
(Small flats)

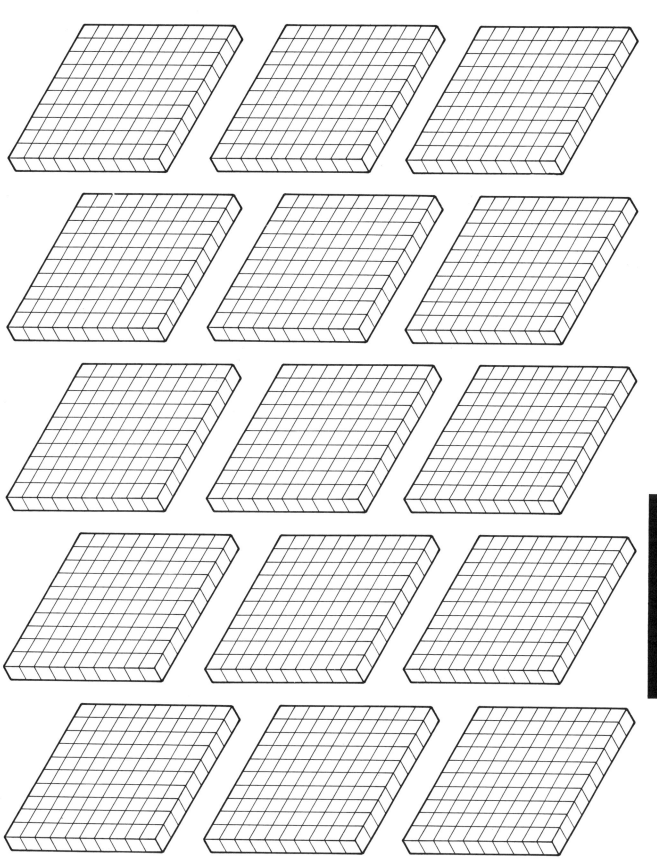

Multi-base Block Cutouts
(Small blocks)

NUMERATION

Bank It or Clear It Gameboard
(Multi-base blocks)
Color in keeping with blocks.
Back and laminate.

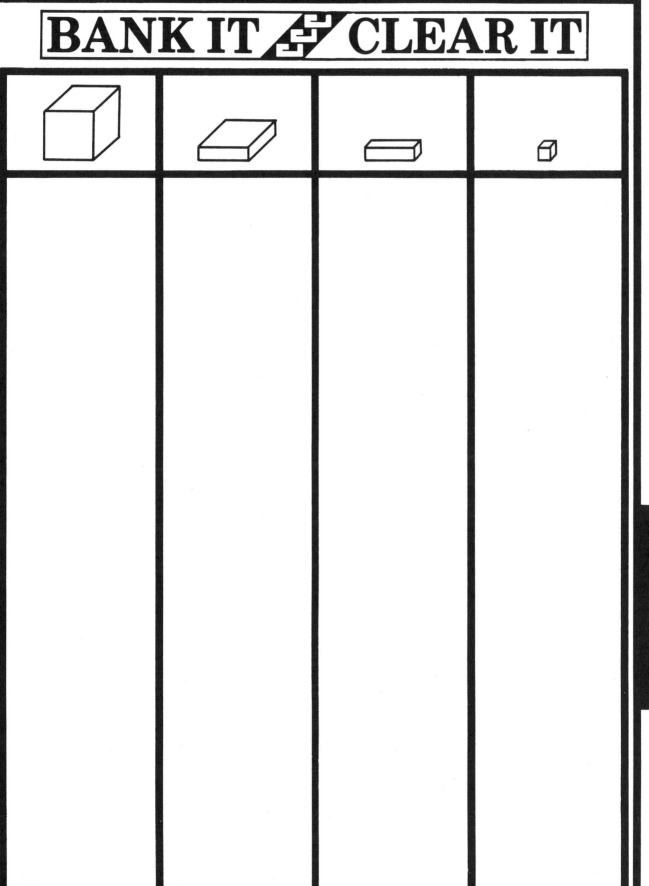

BANK IT / CLEAR IT

NUMERATION

Bank It or Clear It Gameboard
(Counters)
Color in keeping with counters.
Back and laminate.

BANK IT / CLEAR IT

Build a Cube or Break a Cube
Right side **and** left side
Color in keeping with counters.
Back and laminate.

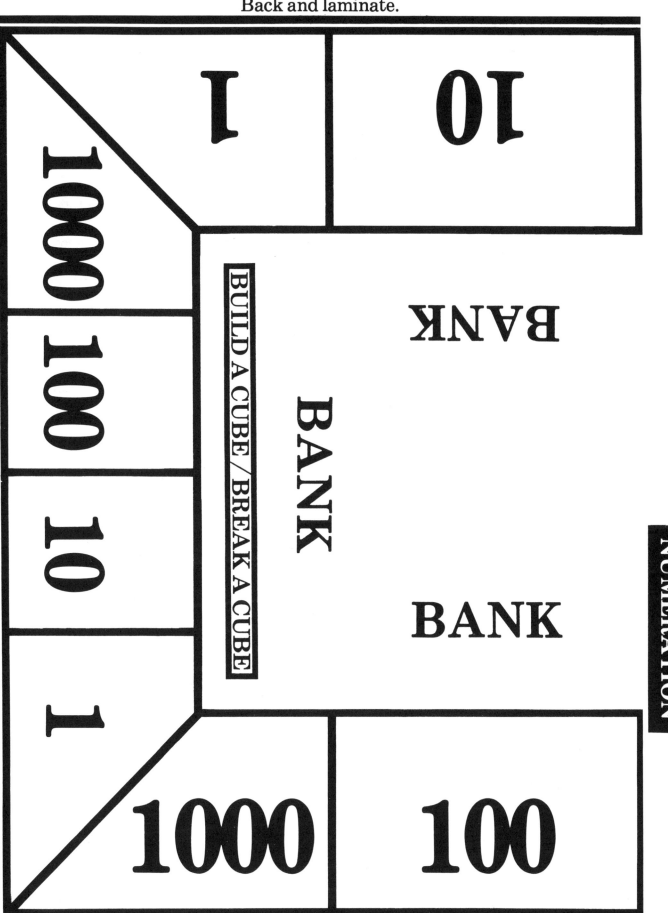

Name...............................

Use the blocks to help you fill in the table. Part of the table has already been filled in as a check for your work.

Number of units	Two land	Three land	Four land	Five land	Six land	Seven land	Eight land	Nine land	Ten land
1	1								
2	10	2							
3	11		3						
4	100					4			
5		12							5
6			12					6	
7		21	13						
8	1000						10		
9		100			13				
10				20				11	

NUMERATION

Addition in Different Lands
(Lands two, three, and six)

Name.......................................

Use the blocks to help you solve the addition problems in the different lands.
Part of the answers to the problems have been given as a check for your work.

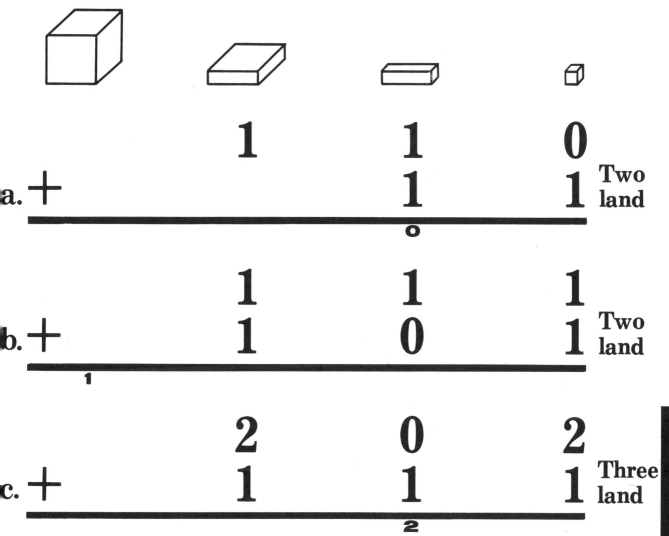

a. $+$ 1   1   0
                    1   1   Two land

		1	1	0	
a.	+		1	1	Two land
			0		

b.	+	1	1	1	Two land
		1	0	1	
	1				

c.	+	2	0	2	Three land
		1	1	1	
			2		

d.	+	1	2	2	1	Three land
			2	1	2	
				1		

e.	+	2	4	3	3	Six land
			3	4	2	
			2			

NUMERATION

MATH GAMES & ACTIVITIES. COPYRIGHT © 1984

103

Addition in different Lands
(Lands seven, eight, and ten)

Name.....................................

Use the blocks to help you solve the addition problems in the different land.
Part of the answers to the problems have been given as a check for your work.

	1	3	0	4	Seve
a. +		4	0	5	land
				2	

	5	6	5	4	Seve
b. +		4	4	3	land
		4			

	1	4	4	3	Eigh
c. + 2		4	2	6	land
	4				

	4	5	4	8	Ten
d. +		8	2	5	land
		3			

	4	4	3	5	Ten
e. + 5		8	7	8	land
	0				

MATH GAMES & ACTIVITIES. COPYRIGHT © 1984

Subtraction in Different Lands
(Lands four, five, and seven)

Name.......................................

Use the blocks to help you solve the subtraction problems in the different lands. Part of the answers to the problems have been given as a check for your work.

	2	**3**	**2**	
a. —		**1**	**3**	Four land
			3	
2	**1**	**3**	**2**	
b. —	**1**	**3**	**3**	Four land
	3			
	3	**4**	**2**	
c. —	**1**	**3**	**3**	Five land
			4	
4	**2**	**4**	**2**	
d. —	**3**	**3**	**3**	Five land
		0		
2	**5**	**2**	**2**	
e. —	**6**	**4**	**3**	Seven land
		4		

NUMERATION

Subtraction in Different Lands
(Lands eight, nine, and ten)

Name.....................................

Use the blocks to help you solve the subtraction problems in the different lands. Part of the answers to the problems have been given as a check for your work.

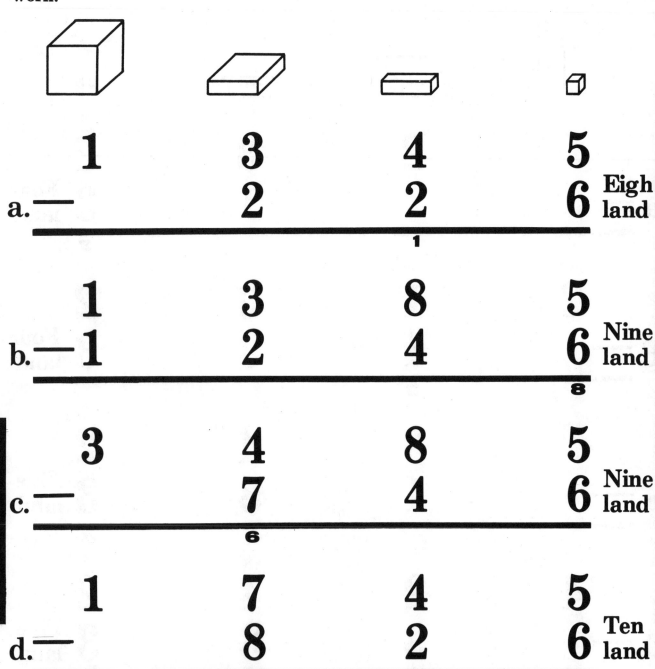

a.	1 — 2	3 2	4 2 (1)	5 6	Eigh land
b.	1 — 1	3 2	8 4	5 6 (8)	Nine land
c.	3 — 4 (6)	4 7	8 4	5 6	Nine land
d.	1 — 7	7 8 (9)	4 2	5 6	Ten land
e.	4 — 4	6 4	1 5 (6)	5 5	Ten land

MATH GAMES & ACTIVITIES. COPYRIGHT © 1984

Ten-Land Task Card
(Card A)
Color in keeping with base ten materials.
Back and laminate.

1. How many?

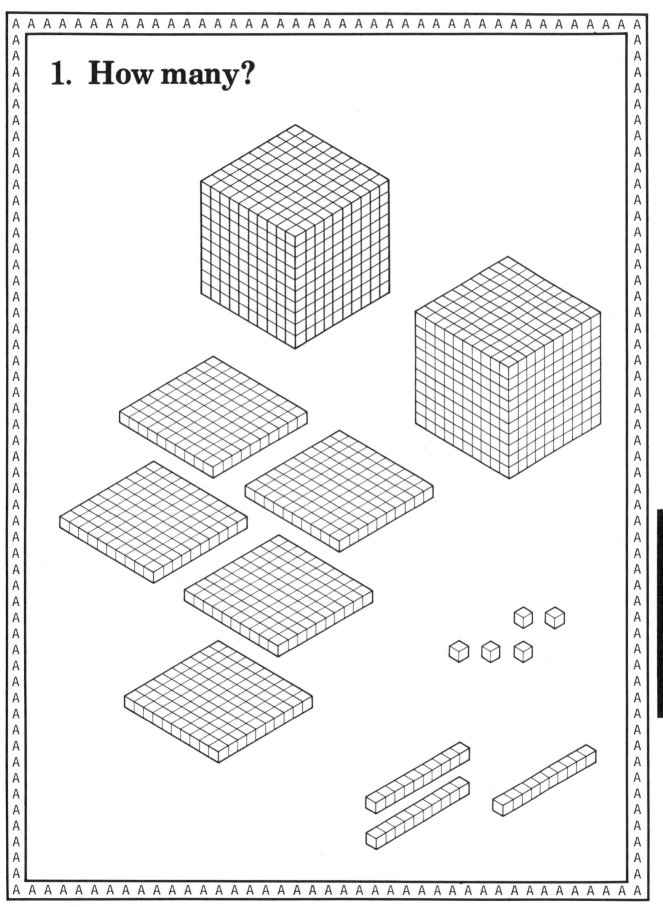

NUMERATION

Ten-Land Task Card
(Card B)
Color in keeping with base ten materials.
Back and laminate.

1. **How many?**

2. **How many for Cards A and B together?**

3. **How many for Card A take away Card B?**

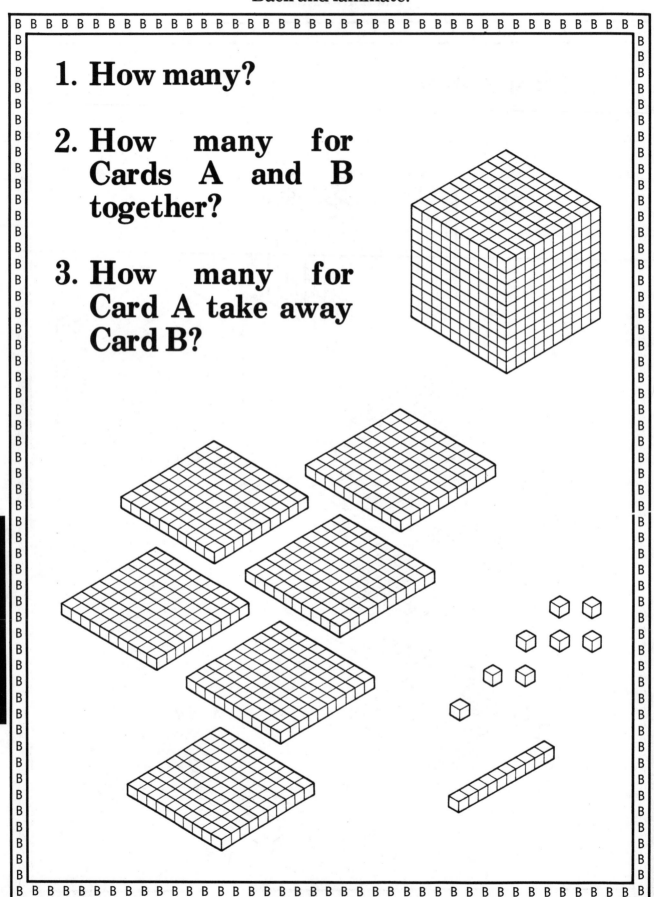

NUMERATION

Name............................

**Color ten jellybeans blue, ten jellybeans red.
How many tens?_____ Ones? _____
Jellybeans? _____**

NUMERATION

Name........................

Color ten fish blue.
Color ten fish red.
Color ten fish yellow.
Color ten fish green.
Color ten fish orange.
How many tens? _____
How many ones? _____
How many fish? _____

NUMERATION

MATH GAMES & ACTIVITIES. COPYRIGHT © 1984

Place Value Rummy Cards
(Page 1 of seven pages)
Back, laminate, and cut along solid lines.

NUMERATION

Three units

Three ones

Three ones

1
1
1

3

3

1000	100						
1000	100	10					
1000	1000	100	100	10	10		
1000	1000	100	100	10	10	1	1
1000	1000	100	100	10	10	1	1
1000	1000	100	100	10	10	1	1

Two longs
Three units

Two tens
Three ones

**Two tens
Three ones**

10 1
10 1
1

23

23

1000	100						
1000	1000	100	100	10	10		
1000	1000	100	100	10	10		
1000	1000	100	100	10	10	1	1
1000	1000	100	100	10	10	1	1
1000	1000	100	100	10	10	1	1
1000	1000	100	100	10	10	1	1

Place Value Rummy Cards
(Page 2 of seven pages)
Back, laminate, and cut along solid lines.

Three longs

Three tens

Three tens

10
10
10

```
1000 1000 1000 1000 1000 1000 1000 1000 1000 1000
100  100  100  100  100  100  100  100  100  100
10   10   10   10   10   10   10   10   10   10
1    1    1    1    1    1    1    1    1    1
```

30

30

Four longs Five units

Four tens Five ones

Four tens Five ones

10 1
10 1
10 1
10 1
1 1000

```
1000 1000 1000 1000 1000 1000 1000 1000 1000 1000
100  100  100  100  100  100  100  100  100  100
10   10   10   10   10   10   10   10   10   10
1    1    1    1    1    1    1    1    1    1
```

45

45

NUMERATION

Place Value Rummy Cards
(Page 3 of seven pages)
Back, laminate, and cut along solid lines.

Five longs
Three units

Five
tens
Three
ones

Five tens
Three ones

53

53

10 1
10 1
10 1
10 1

1000 1000 1000 1000 1000 1000 1000 1000 1000
100 100 100 100 100 100 100 100 100
10 10 10 10 10 10 10 10 10
1 1 1 1 1 1 1 1 1

One
flat
Two
units

One
hundred
Two
ones

One hundred
Two ones

102

102

100 1
1

1000 1000 1000 1000 1000 1000 1000 1000 1000 1000
100 100 100 100 100 100 100 100 100 100
10 10 10 10 10 10 10 10 10 10
1 1 1 1 1 1 1 1 1 1

Place Value Rummy Cards
(Page 4 of seven pages)
Back, laminate, and cut along solid lines.

NUMERATION

Two flats
One unit

100 1
100

```
1000  1000  1000  1000  1000  1000  1000  1000  1000
1000  1000  1000  1000  1000  1000  1000  1000  1000
100   100   100   100   100   100   100   100   [100]
                                                10   10   10   10   10   10   10   10   10
                                                1    1    1    1    1    1    1    1    [1]
```

Two hundreds
One one

201

201

Two hundreds
One one

**Two hundreds
One one**

Three flats

100
100
100

```
1000  1000  1000  1000  1000  1000  1000  1000  1000
1000  1000  1000  1000  1000  1000  1000  1000  1000
100   100   100   100   100   100   [100]  100   100
10    10    10    10    10    10    10    10    10
1     1     1     1     1     1     1     1     1
```

Three hundreds

300

300

Three hundreds

**Three hun-
dreds**

Place Value Rummy Cards
(Page 5 of seven pages)
Back, laminate, and cut along solid lines.

NUMERATION

One
flat
Two
longs
Three
units

One
hundred
Two
tens
Three
ones

**One hundred
Two tens
Three ones**

100 10 1
10 1
1

1000
1000 1000
1000 1000
1000 1000 100
1000 1000 100
1000 1000 100 (100)
1000 1000 100 10
1000 1000 100 10
1000 1000 100 10 (10)
1000 1000 100 10 (10)
1000 100 10 1
1000 100 10 1 (1)
1000 100 10 1 1

123

123

One
flat
One
long
One
unit

One
hundred
One
ten
One
one

**One hundred
One ten
One one**

100 10 1

1000 1000
1000 1000
1000 1000 100
1000 1000 100
1000 1000 100 (100)
1000 1000 100 10
1000 1000 100 10
1000 1000 100 10 (10)
1000 1000 100 10 10
1000 100 10 1
1000 100 10 1 (1)

111

111

Place Value Rummy Cards
(Page 6 of seven pages)
Back, laminate, and cut along solid lines.

NUMERATION

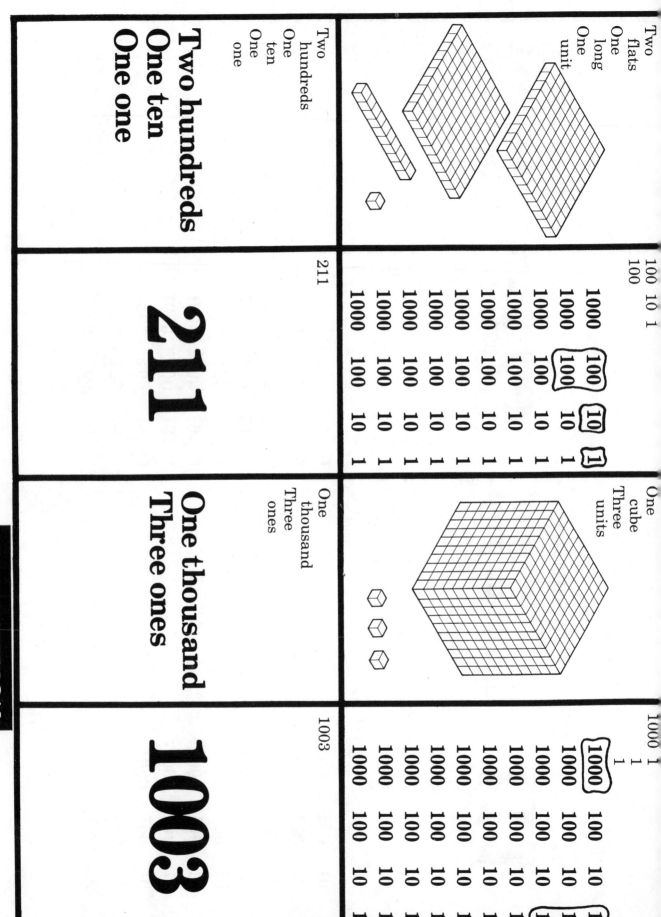

Two
flats
One
long
One
unit

100 10 1

Two
hundreds
One
ten
One
one

211

```
1000 1000 1000 1000 1000 1000 100  100
1000 1000 1000 1000 1000 1000 1000 100
1000 1000 1000 1000 1000 1000 1000 10
                                    1
```

211

**Two hundreds
One ten
One one**

One
cube
Three
units

One
thousand
Three
ones

1000 1

1003

```
1000 1000 1000 1000 1000 1000 1000 1000
1000 1000 1000 1000 1000 1000 1000 100
1000 1000 1000 1000 1000 1000 1000 100
100  100  100  100  100  100  100  100
10   10   10   10   10   10   10   10
1    1    1    1    1    1    1    1
```

1003

**One thousand
Three ones**

Place Value Rummy Cards
(Page 7 of seven pages)
Back, laminate, and cut along solid lines.

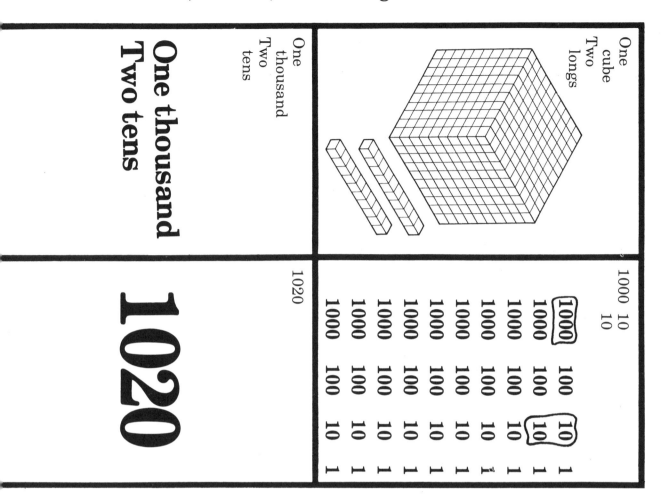

One
cube
Two
longs

One
thousand
Two
tens

**One thousand
Two tens**

1020

1020

1000 10
       10

$$\boxed{1000}$$ 1000 1000 1000 1000 1000 1000 1000 1000 1000 1000
100 100 100 100 100 100 100 100 100 100
$$\boxed{10}$$ 10 10 10 10 10 10 10 10 10
1 1 1 1 1 1 1 1 1 1

NUMERATION

START 1000

1

10

100

1000

One's place

Ten's place

Hun-dred's place

Sprained an-kle. Go back one space.

Thou-sand's place

Ten's place

Hun-dred's place

10

One's place

1000

Slid downhill. Go forward three spaces.

100

Saw bear. Go back three spaces.

1

10

NUMERATION

118

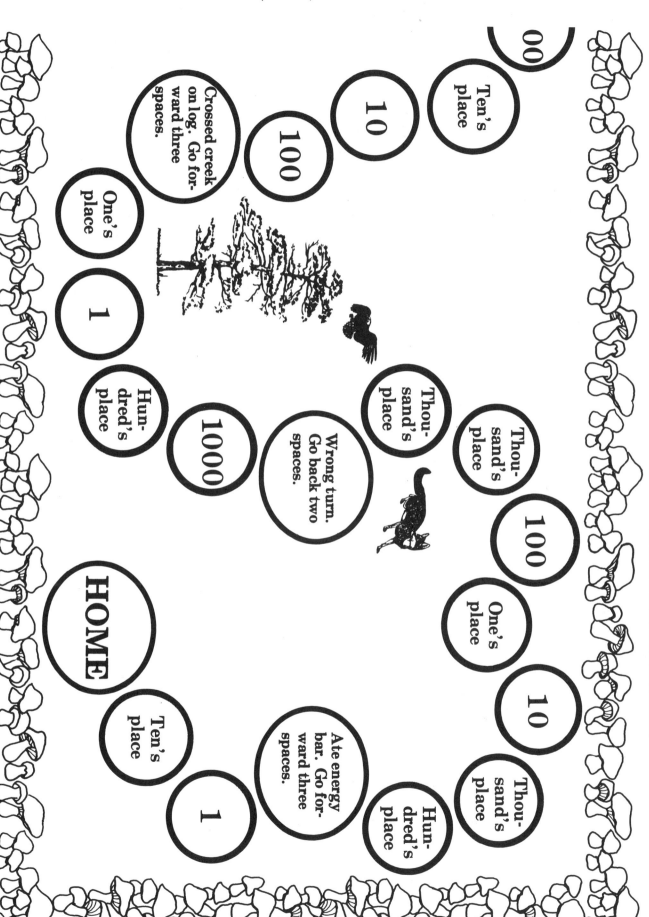

NUMERATION

Digit Cards
(Page 1 of two pages)
Back, laminate, and cut along solid lines.

DIGIT CARD	DIGIT CARD	DIGIT CARD	DIGIT CARD
4051	**3816**	**2390**	**5274**
DIGIT CARD	DIGIT CARD	DIGIT CARD	DIGIT CARD
8423	**6502**	**9147**	**1035**
DIGIT CARD	DIGIT CARD	DIGIT CARD	DIGIT CARD
1508	**5134**	**9320**	**7642**
DIGIT CARD	DIGIT CARD	DIGIT CARD	DIGIT CARD
2463	**9208**	**4715**	**3051**
DIGIT CARD	DIGIT CARD	DIGIT CARD	DIGIT CARD
4501	**8315**	**2430**	**6279**

NUMERATION

Digit Cards
(Page 2 of two pages)
Back, laminate, and cut along solid lines.

DIGIT CARD	DIGIT CARD	DIGIT CARD	DIGIT CARD
3295	**2460**	**8417**	**5031**
DIGIT CARD	DIGIT CARD	DIGIT CARD	DIGIT CARD
1084	**7153**	**4302**	**5629**
DIGIT CARD	DIGIT CARD	DIGIT CARD	DIGIT CARD
9523	**6402**	**1745**	**3180**
DIGIT CARD	DIGIT CARD	DIGIT CARD	DIGIT CARD
5810	**4371**	**2036**	**9254**
DIGIT CARD	DIGIT CARD	DIGIT CARD	DIGIT CARD
9235	**8042**	**7416**	**5103**

NUMERATION

Ollie Octopus Card
Card 1
Color, back, and laminate.

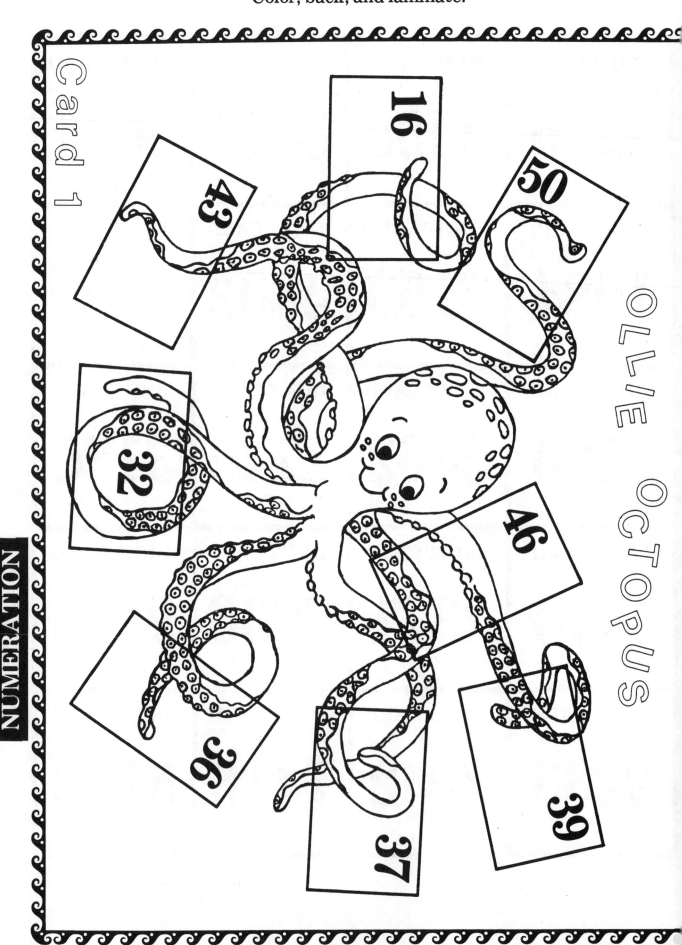

OLLIE OCTOPUS

NUMERATION

Ollie Octopus Card
Card 2
Color, back, and laminate.

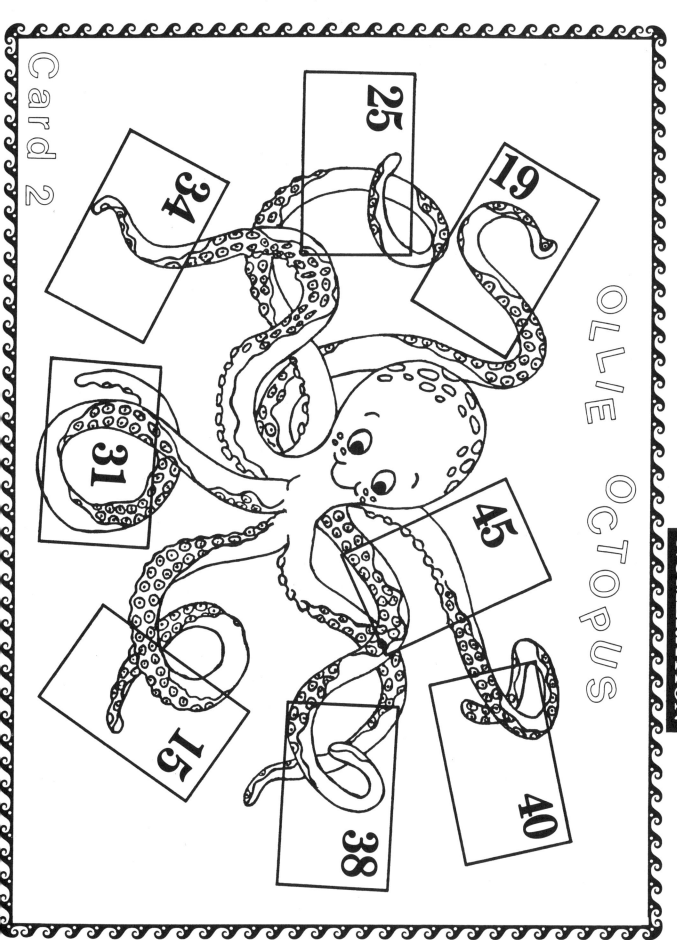

Card 2

OLLIE OCTOPUS

Ollie Octopus Card
Card 3
Color, back, and laminate.

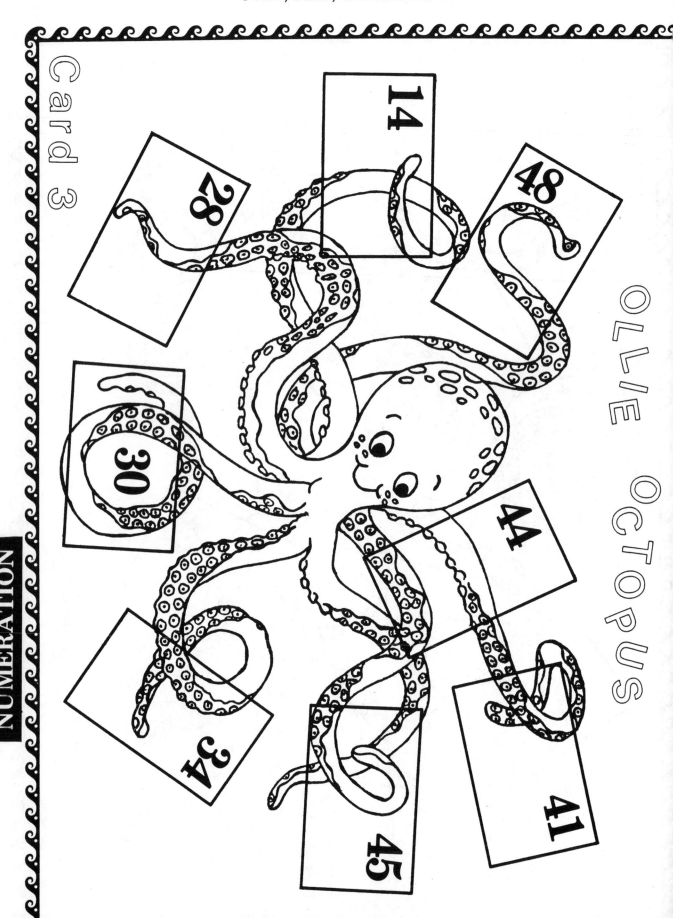

NUMERATION

OLLIE OCTOPUS

Ollie Octopus Card
Card 4
Color, back, and laminate.

Card 4

23

47

27

29

OLLIE OCTOPUS

43

33

39

12

Ollie Octopus Card
(Blank)

Card

NUMERATION

OLLIE OCTOPUS

Ollie Octopus Tags
(Numeration)
Back, laminate, and cut along solid lines.

OLLIE OCTOPUS 5 tens	OLLIE OCTOPUS 40 + 6	OLLIE OCTOPUS 3 tens + 9 ones	OLLIE OCTOPUS 20 + 17
OLLIE OCTOPUS 36 ones	OLLIE OCTOPUS 30 + 2	OLLIE OCTOPUS 4 tens + 3 ones	OLLIE OCTOPUS 10 + 6
OLLIE OCTOPUS 10 + 9	OLLIE OCTOPUS 4 tens + 5 ones	OLLIE OCTOPUS 4 tens	OLLIE OCTOPUS 20 + 18
OLLIE OCTOPUS 1 ten + 5 ones	OLLIE OCTOPUS 20 + 11	OLLIE OCTOPUS 30 + 4	OLLIE OCTOPUS 25 ones
OLLIE OCTOPUS 40 + 8	OLLIE OCTOPUS 4 tens + 4 ones	OLLIE OCTOPUS 30 + 11	OLLIE OCTOPUS 30 + 15
OLLIE OCTOPUS 3 tens + 4 ones	OLLIE OCTOPUS 20 + 10	OLLIE OCTOPUS 2 tens + 8 ones	OLLIE OCTOPUS 10 + 4
OLLIE OCTOPUS 40 + 7	OLLIE OCTOPUS 30 + 13	OLLIE OCTOPUS 10 + 2	OLLIE OCTOPUS 30 + 9
OLLIE OCTOPUS 20 + 13	OLLIE OCTOPUS 2 tens + 9 ones	OLLIE OCTOPUS 10 + 17	OLLIE OCTOPUS 2 tens + 3 ones
OLLIE OCTOPUS 30 + 4	OLLIE OCTOPUS 30 + 9	OLLIE OCTOPUS 40 + 3	OLLIE OCTOPUS 40 + 5

NUMERATION

Addition Facts Table

+	0	1	2	3	4	5	6	7	8	9
0	0	1	2	3	4	5	6	7	8	9
1	1	2	3	4	5	6	7	8	9	10
2	2	3	4	5	6	7	8	9	10	11
3	3	4	5	6	7	8	9	10	11	12
4	4	5	6	7	8	9	10	11	12	13
5	5	6	7	8	9	10	11	12	13	14
6	6	7	8	9	10	11	12	13	14	15
7	7	8	9	10	11	12	13	14	15	16
8	8	9	10	11	12	13	14	15	16	17
9	9	10	11	12	13	14	15	16	17	18

Balance Beam Cutout
Back, laminate, and punch out holes.

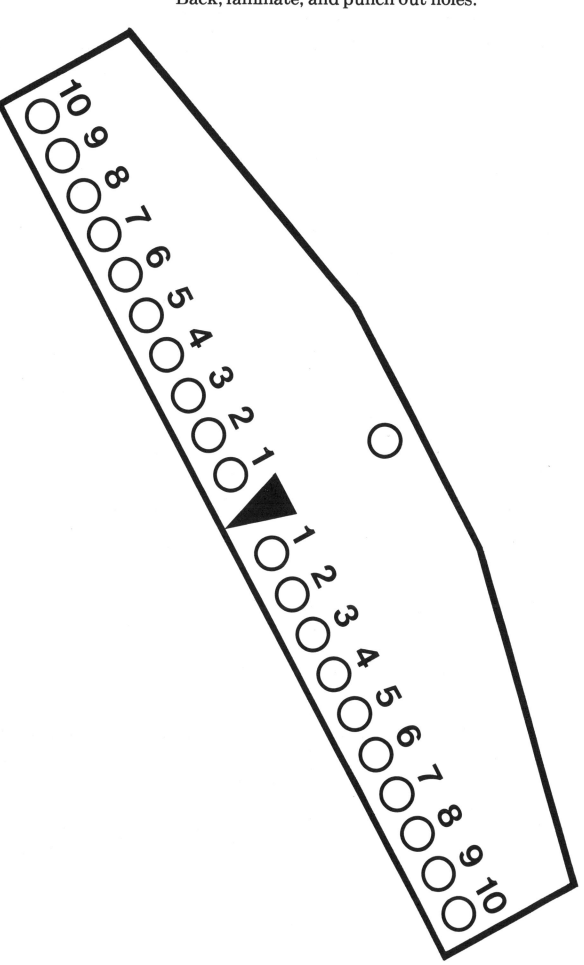

Name......................

a.

___ + ___ = ___

___ + ___ = ___

___ + ___ = ___

b.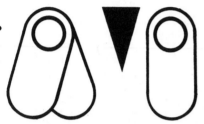

___ × ___ = ___

___ × ___ = ___

___ × ___ = ___

c.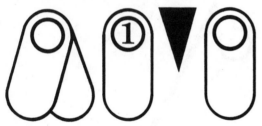

___ × ___ + 1 = ___

___ × ___ + 1 = ___

___ × ___ + 1 = ___

d.

___ × ___ = ___ + ___

___ × ___ = ___ + ___

___ × ___ = ___ + ___

e.

___ × ___ = ___ × ___ + ___ × ___

___ × ___ = ___ × ___ + ___ × ___

___ × ___ = ___ × ___ + ___ × ___

Balance Beam Worksheet
(Second of three)

Name......................

a.

___ + ___ = ___ + ___

___ + ___ = ___ + ___

___ + ___ = ___ + ___

b.

___ + ___ + ___ = ___

___ + ___ + ___ = ___

___ + ___ + ___ = ___

c.

___ < ___   ___ > ___

___ < ___   ___ > ___

___ < ___   ___ > ___

d.

___ + ___ + ___ = ___ × ___

___ + ___ + ___ = ___ × ___

___ + ___ + ___ = ___ × ___

e.

___ + ___ + ___ = ___ + ___

___ + ___ + ___ = ___ + ___

___ + ___ + ___ = ___ + ___

f.

___ × ___ = ___ × ___

___ × ___ = ___ × ___

___ × ___ = ___ × ___

ADDITION FACTS

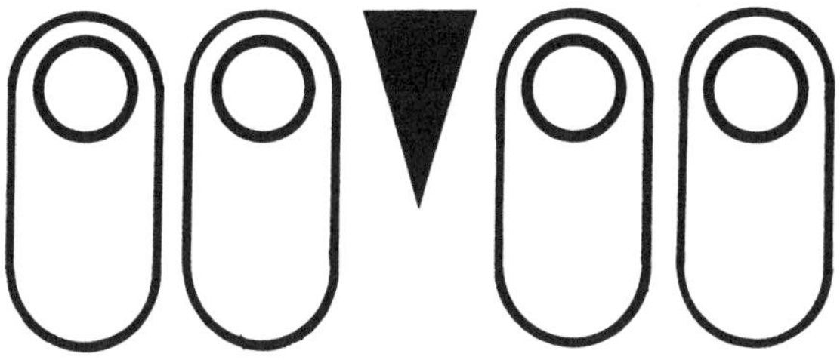

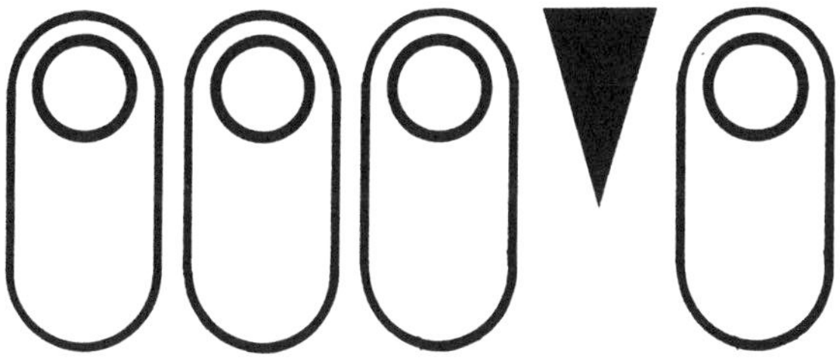

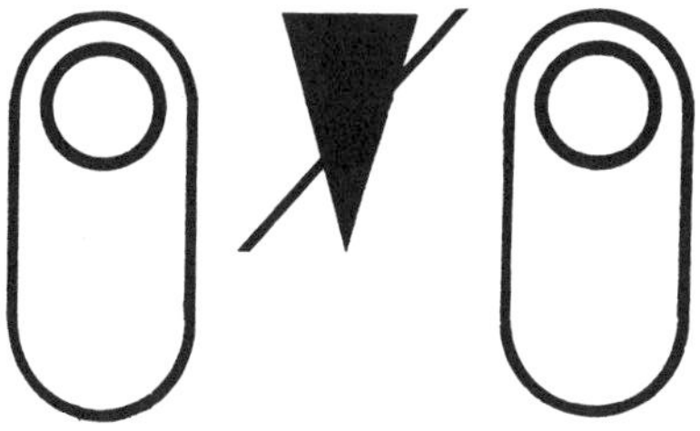

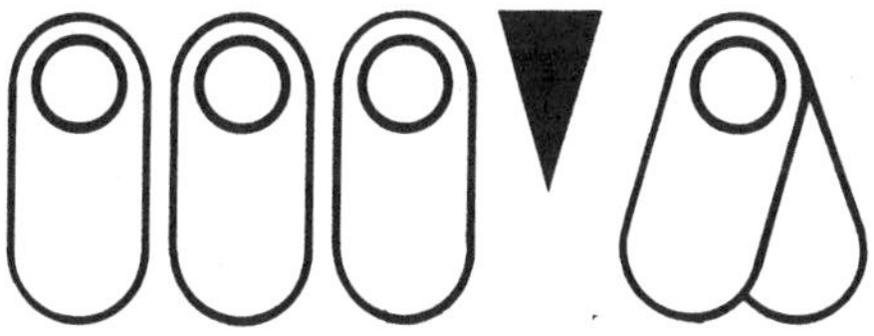

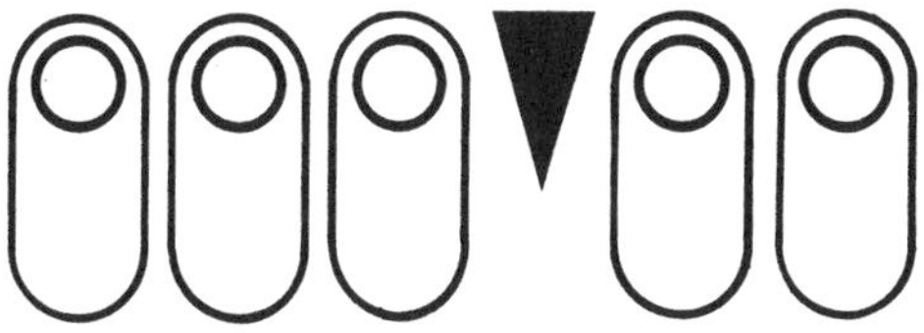

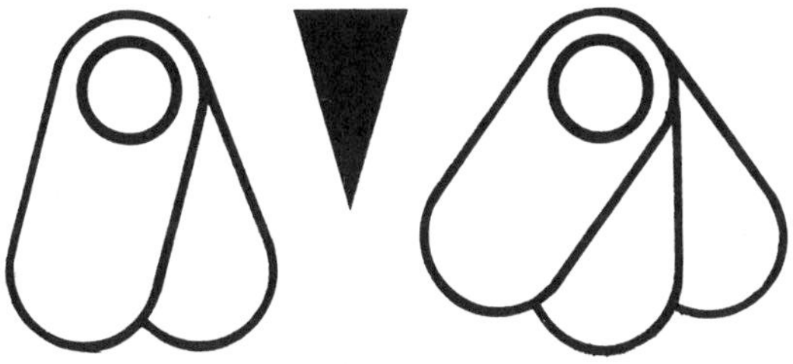

Name........................

a.

___ × ___ = ___ × ___

___ × ___ = ___ × ___

b.

___ × ___ = ___ × ___

___ × ___ = ___ × ___

c.

___ × ___ = ___ + ___

___ × ___ = ___ + ___

___ × ___ = ___ + ___

d.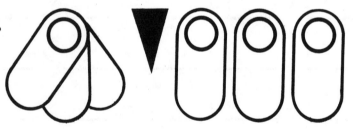

___ × ___ = ___ + ___ + ___

___ × ___ = ___ + ___ + ___

___ × ___ = ___ + ___ + ___

e.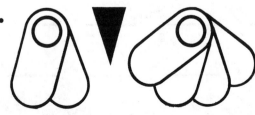

___ × ___ = ___ × ___

___ × ___ = ___ × ___

___ × ___ = ___ × ___

Frog
(Addition facts)

Name..

Color the frog. Four = green, 6 = yellow, 8 = brown, and 10 = red.

Name...

Color the frog. Four = green, 6 = yellow, 8 = brown, and 10 = red.

MATH GAMES & ACTIVITIES. COPYRIGHT © 1984

Puppies
(Addition facts)

Name...

Which puppies are mine?

$1 + 9$

10

$6 + 6$

$8 + 2$

$7 + 3$

$9 + 3$

$4 + 5$

$3 + 7$

$5 + 5$

$4 + 6$

$3 + 8$

$2 + 7$

ADDITION FACTS

Puppies
(Blank)

Name...

MATH GAMES & ACTIVITIES. COPYRIGHT © 1984

Name...................................

Find the sums. Then join the dots in order. When finished, you will have a lovely picture.

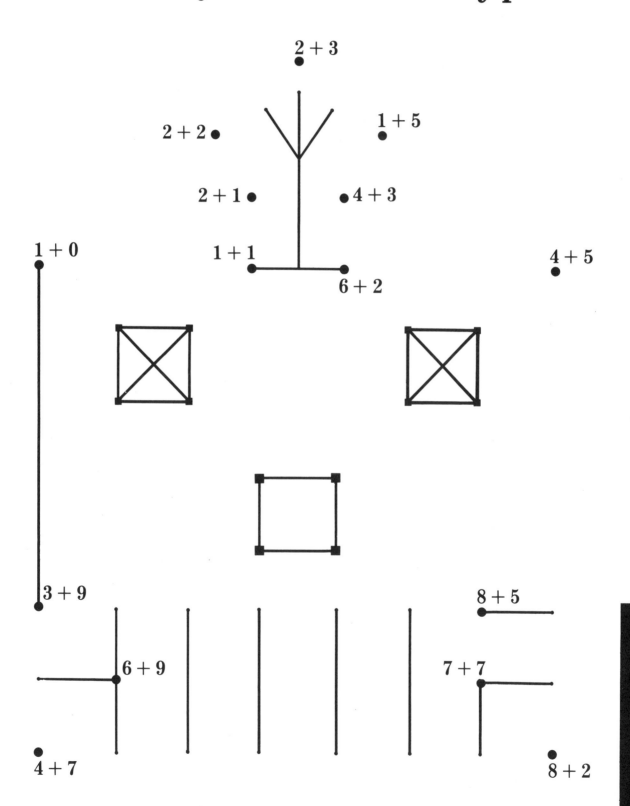

2 + 3

2 + 2 •

1 + 5

2 + 1 •

• 4 + 3

1 + 0

1 + 1

4 + 5

6 + 2

3 + 9

8 + 5

6 + 9

7 + 7

4 + 7

8 + 2

ADDITION FACTS

Name............................

Find the shortest path from the castle to the church.

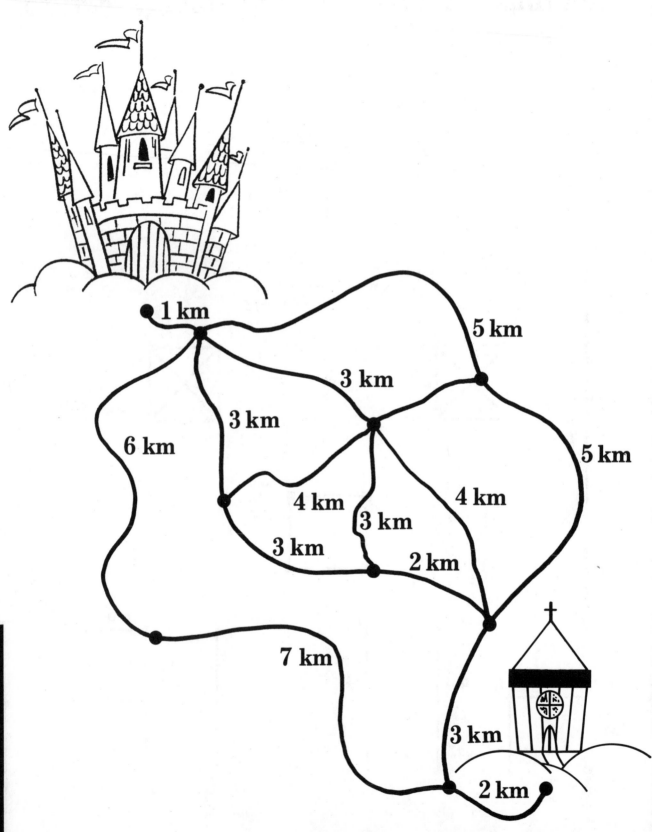

     MATH GAMES & ACTIVITIES. COPYRIGHT © 1984

Name.................................

How old is the man? Add the numbers to find out. Answer _____

Staircase Gameboard
Back and laminate.
(Do **not** color the staircase.)

ROD COUNTER

STAIRCASE

Boat Gameboard
Back and laminate.
(Do not color the boat.)

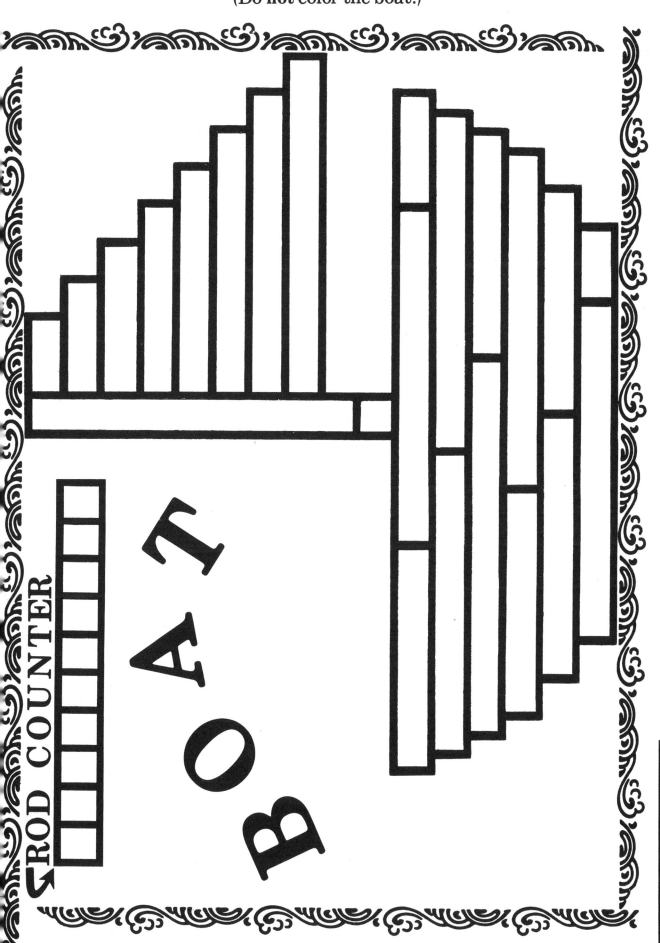

ROD COUNTER

BOAT

ADDITION FACTS

Horse Gameboard
Back and laminate.
(Do **not** color the horse.)

HORSE

ROD COUNTERS

Engine Gameboard
Back and laminate.
(Do **not** color the engine.)

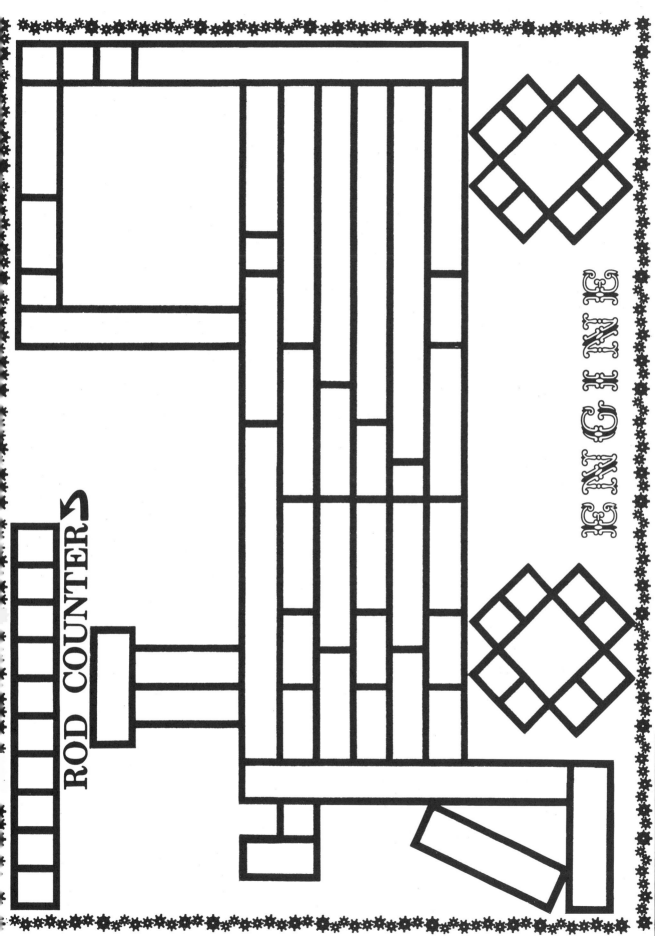

ENGINE

ROD COUNTERS

ADDITION FACTS

143

Shooting Gallery Gameboard
(Addition facts)
Color, back, and laminate.

4 + 2

2 + 2

2 + 3

1 + 0

2 + 1

1 + 1

SHOOTING GALLERY

1 + 1

1 + 2

2 + 0

3 + 3

1 + 0

3 + 1

SHOOTING GALLERY

ADDITION FACTS

Addition Facts Rummy Cards
(Page 1 of seven pages)
Color shapes, back, and laminate.
Cut along solid lines.

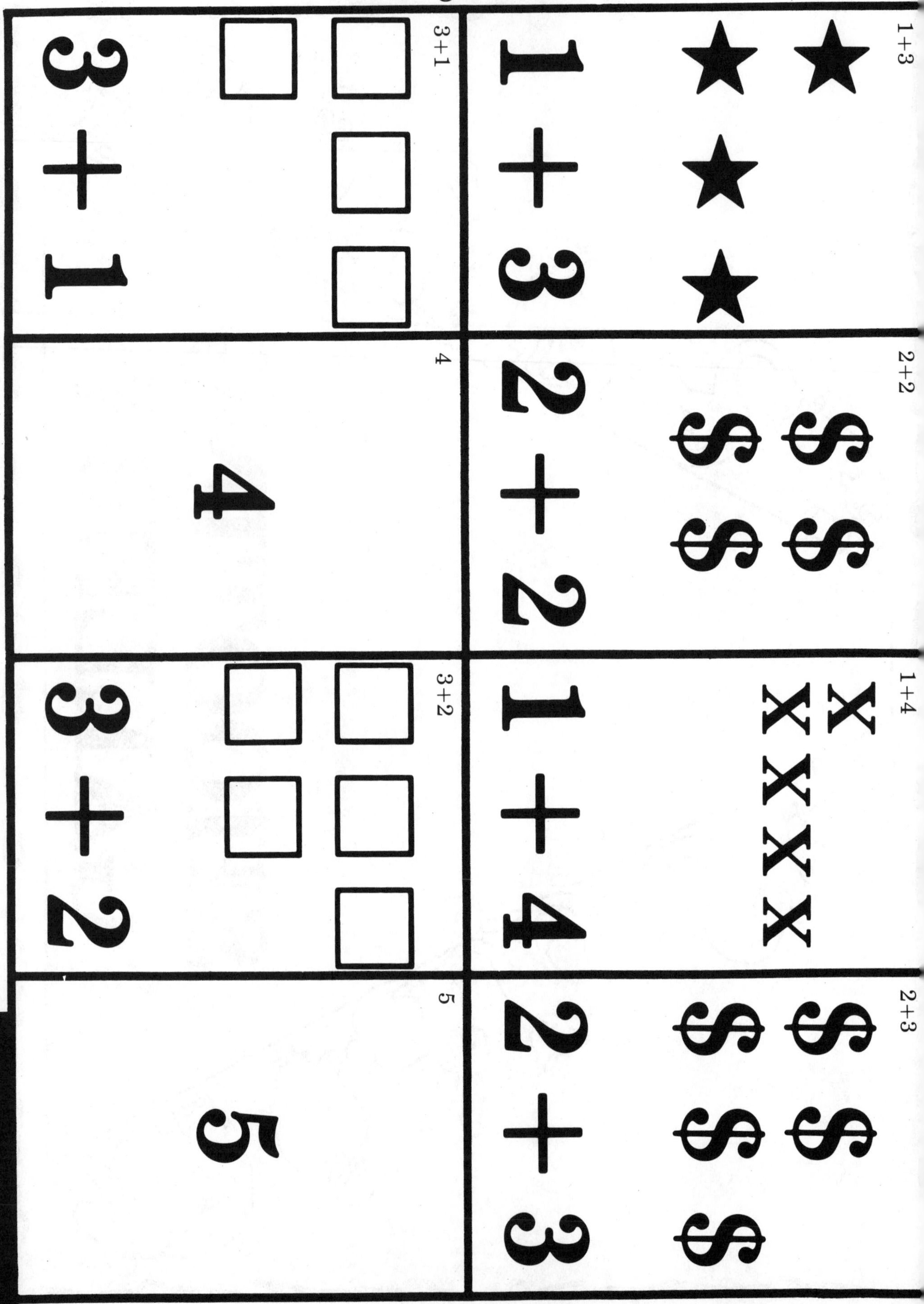

1+3

1 + 3 ★ ★ ★ ★ ★

3+1

3 + 1 □ □ □ □

2+2

2 + 2 $ $ $ $

4

4

1+4

1 + 4 X X X X X

3+2

3 + 2 □ □ □ □ □

2+3

2 + 3 $ $ $ $ $

5

5

Addition Facts Rummy Cards
(Page 2 of seven pages)
Color shapes, back, and laminate.
Cut along solid lines.

1+5

$ $
$ $
$ $
$ $
$ $

1 + 5

3+3

□ □
□ □
□ □

3 + 3

2+4

X X
X X X
X X X

2 + 4

6

6

1+6

✓ ✓
✓ ✓ ✓
✓ ✓

1 + 6

3+4

1 1 1
1 1 1
1 1 1
1

3 + 4

2+5

$ $ $
$ $ $
$ $
$ $
$

2 + 5

7

7

ADDITION FACTS

Addition Facts Rummy Cards
(Page 3 of seven pages)
Color shapes, back, and laminate.
Cut along solid lines.

2+6

★ ★
★ ★
★
★
★
★

4+4

$ $
$ $
$ $
$ $

4 + 4

2 + 6

3+5

ᴔ ᴔ
ᴔ ᴔ
ᴔ
ᴔ

3 + 5

8

8

1+8

* *
*
*
*
*
*
*
*

4+5

$ $
$ $
$ $
$ $
$

4 + 5

1 + 8

3+6

ʅ ʅ
ʅ ʅ
ʅ ʅ
ʅ
ʅ

9

9

3 + 6

Addition Facts Rummy Cards
(Page 4 of seven pages)
Color shapes, back, and laminate.
Cut along solid lines.

3+7

$$3 + 7$$

5+5

$$5 + 5$$

4+6

$$4 + 6$$

10

10

3+8

$$3 + 8$$

5+6

$$5 + 6$$

4+7

$$4 + 7$$

11

11

ADDITION FACTS

Addition Facts Rummy Cards
(Page 5 of seven pages)
Color shapes, back, and laminate.
Cut along solid lines.

3+9

□ □
□ □
□
□
□
□
□
□

6+6

/ /
/ /
/ /
/ /
/ /
/ /

6 + 6

5+7

★ ★
★ ★
★ ★
★ ★
★ ★
★
★

12

12

4+9

/ /
/ /
/ /
/
/

3 + 9

5 + 7

6+7

* *
* *
* *
* *
* *
*

6 + 7

5+8

1 1 1 1
1 1 1 1
1 1 1 1
1 1 1 1

4 + 9

13

13

5 + 8

Addition Facts Rummy Cards
(Page 6 of seven pages)
Color shapes, back, and laminate.
Cut along solid lines.

5+9

□ □
□ □
□ □
□ □
□ □
□
□

5 + 9

7+7

★ ★
★ ★
★ ★
★ ★
★ ★
★ ★
★ ★
★ ★

7 + 7

6+8

1 1 1 1 1
1 1 1 1
1 1 1 1

6 + 8

14

14

6+9

* * *
* * *
* * *
* * *
* * *
*
*

6 + 9

7+8

x x x x x x
x x x x x x
x x x x x
x x x

7 + 8

8+7

8 + 7

9+6

\ \ \
\ \ \
\ \ \
\ \ \
\ \
\ \
\

9 + 6

15

15

ADDITION FACTS

Addition Facts Rummy Cards
(Page 7 of seven paes)
Color shapes, back, and laminate.
Cut along solid lines.

9 + 7

7 + 9

9+7

7+9

8 + 8

16

16

8+8

Name.............

Color the picture. Three = brown, 5 = yellow, 7 = red, and 9 = green.

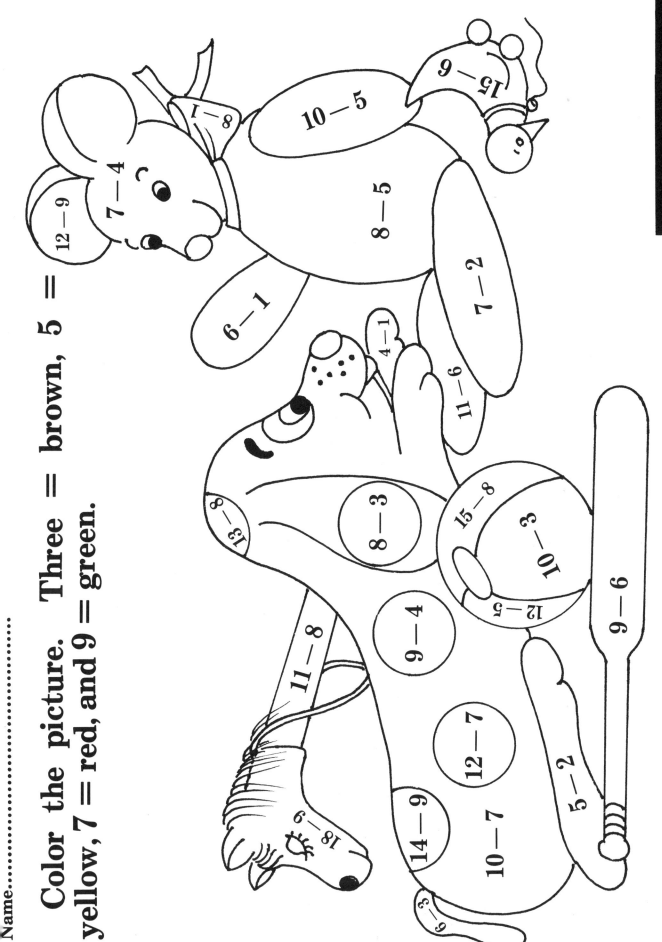

SUBTRACTION FACTS

Toys
(Blank)

Name.......................

Color the picture. Three = brown, 5 = yellow, 7 = red, and 9 = green.

154

Balloon Pop Gameboard
(Subtraction facts)
Color, back, and laminate.

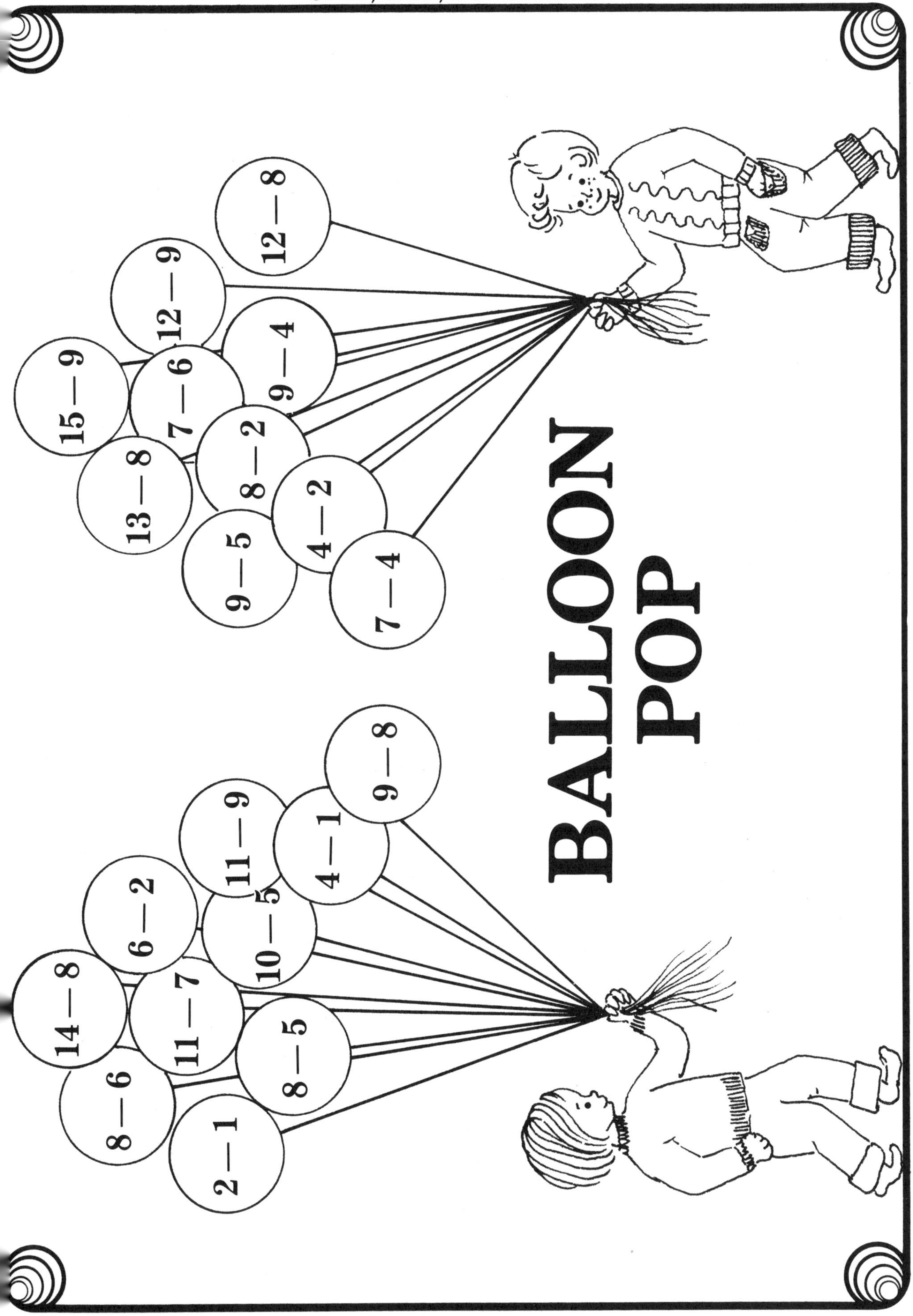
BALLOON
POP
12—8
12—9
9—4
15—9
7—6
8—2
13—8
4—2
9—5
7—4
9—8
11—9
4—1
6—2
10—5
14—8
11—7
8—5
8—6
2—1

Balloon Pop Gameboard
(Blank)

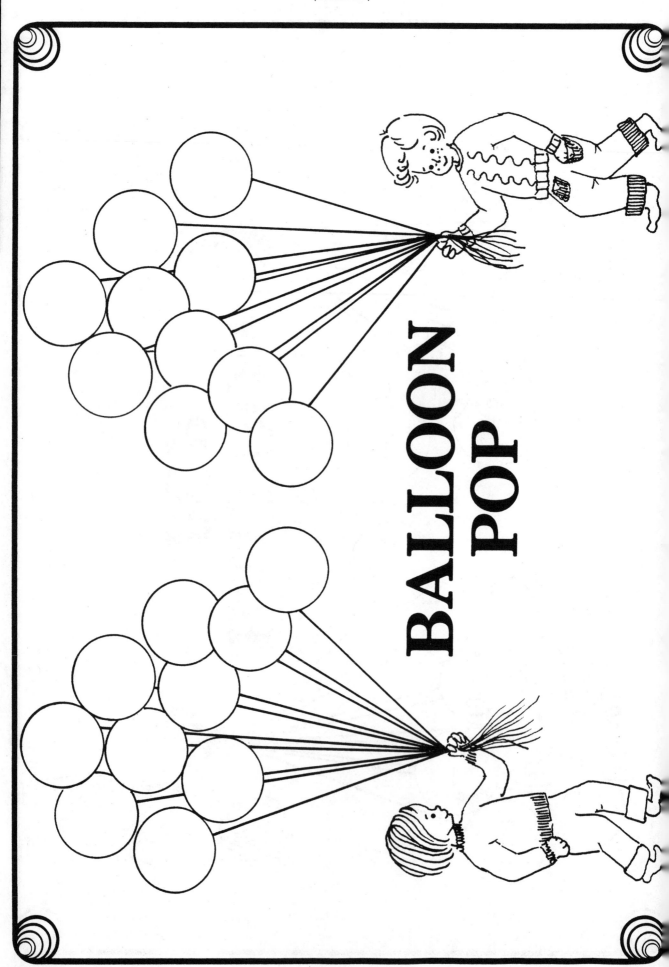

BALLOON POP

Seesaw
(Related facts)

Name.......................

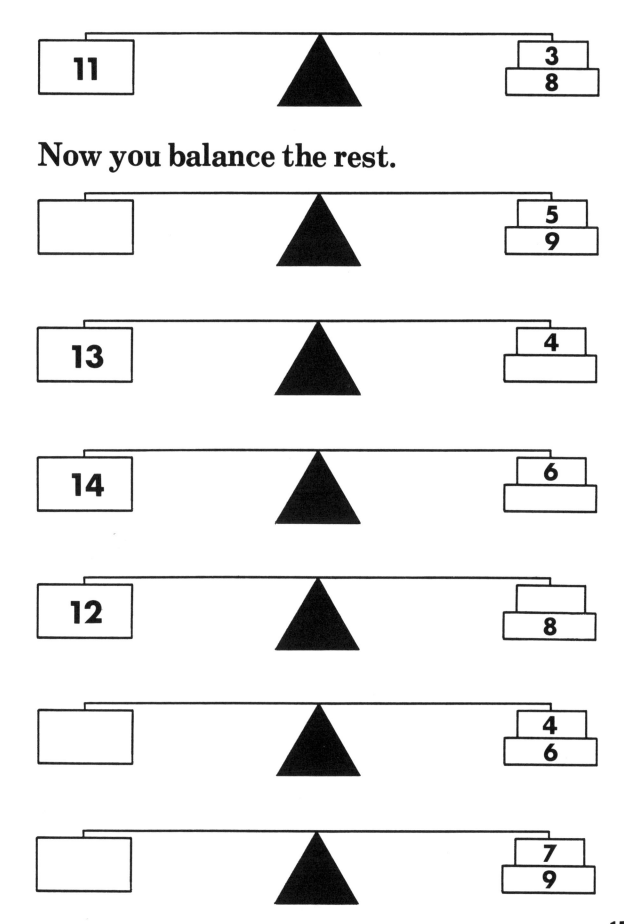

Now you balance the rest.

ADDITION AND SUBTRACTION FACTS

11 | 3 / 8

5 / 9

13 | 4

14 | 6

12 | 8

4 / 6

7 / 9

Name. . . . .

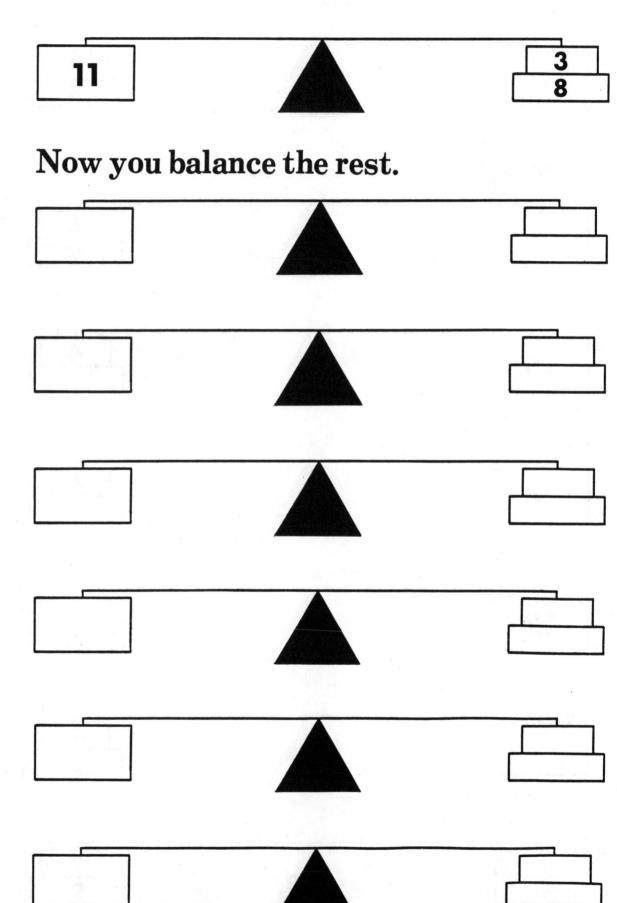

Now you balance the rest.

ADDITION AND SUBTRACTION FACTS

Name..

Write as many number sentences for each lady bug as you can.

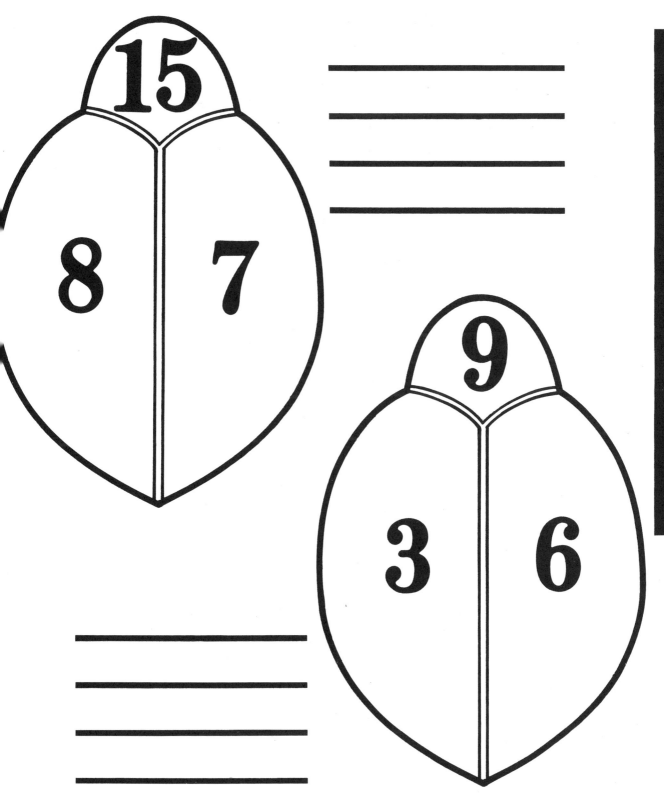

Name...

Write as many number sentences for each lady bug as you can.

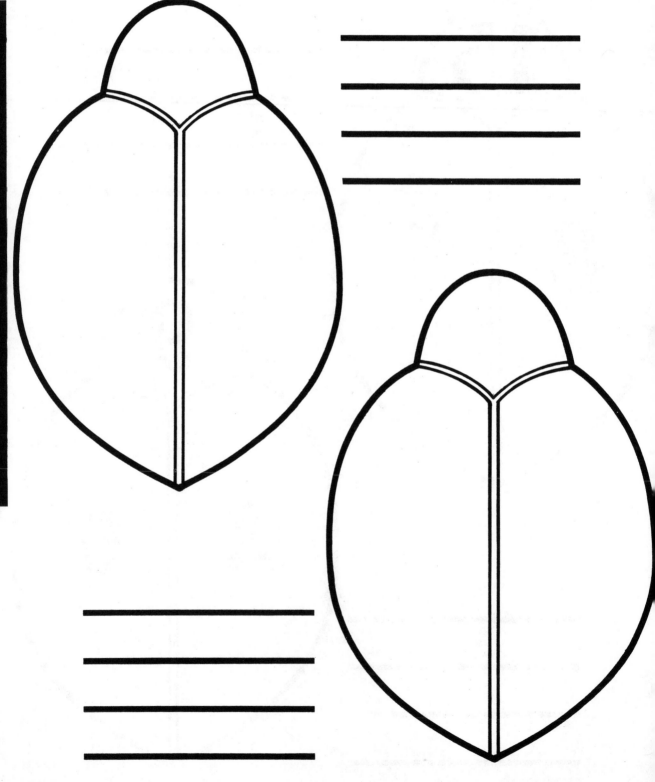

ame..............................

5

$5 + 0 = 5$	$5 - 0 = 5$
$4 + 1 = 5$	$5 - 1 = 4$
$3 + 2 = 5$	$5 - 2 = 3$
$2 + 3 = 5$	$5 - 3 = 2$
$1 + 4 = 5$	$5 - 4 = 1$
$0 + 5 = 5$	$5 - 5 = 0$

Now you fill in the rest.

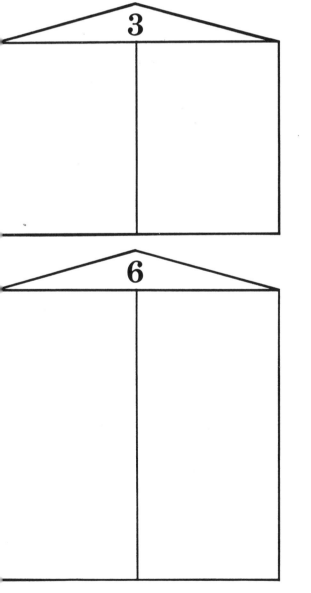

3

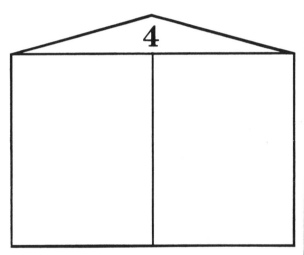

4

6

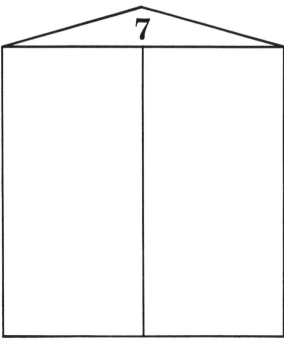

7

Dinosaur Eggs
(Addition and subtraction facts)

Three broken dinosaur eggs. Cut them ou and mend them. Use the numbers to help you.

Dinosaur Eggs
(Blank)

Three broken dinosaur eggs. Cut them out and mend them. Use the numbers to help you.

Computasnake
(Addition and subtraction facts)

Name..............................

Find the answers to each part of the snake. Match the answers with the houses. Color the houses the snake has been to.

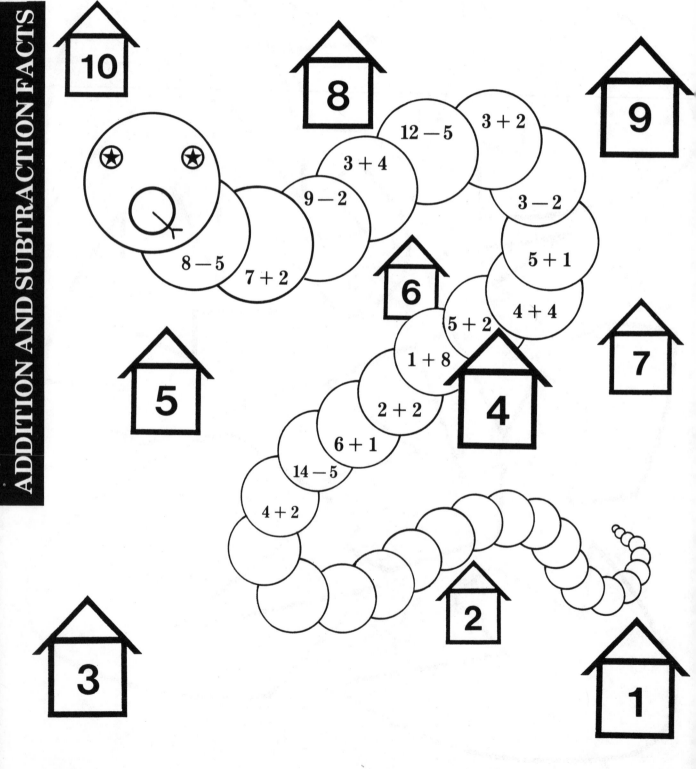

Name...................................

Find the answers to each part of the snake. Match the answers with the houses. Color the houses the snake has been to.

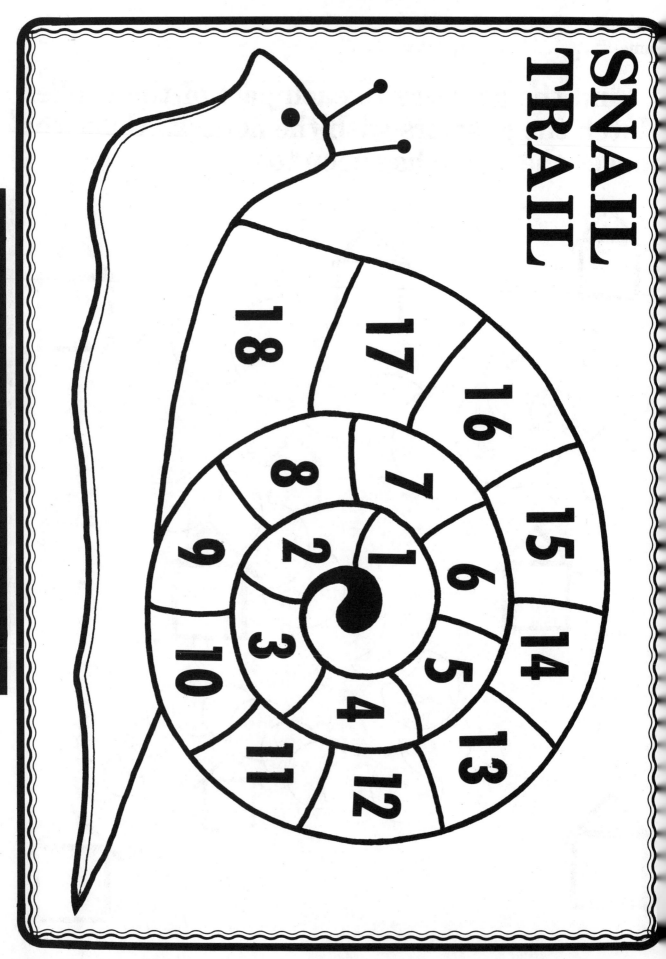

SNAIL TRAIL

ADDITION AND SUBTRACTION FACTS

Addition and Subtraction Facts Dominoes
(Page 1 of six pages)
Back, laminate, and cut along dotted lines.

ADDITION AND SUBTRACTION FACTS

ZERO	□	4 — 4	5 — 4
1 — 1	7 — 4	0 + 0	3 + 1
2 — 2	14 — 8	5 — 5	14 — 7
3 — 3	13 — 4	1	ONE
1 + 0	11 — 8	4 — 3	4 + 0

Addition and Subtraction Facts Dominoes
(Page 2 of six pages)
Back, laminate, and cut along dotted lines.

ADDITION AND SUBTRACTION FACTS

10 — 7	2 + 4	9 — 6	2 + 5
12 — 9	3 + 6	FOUR	4
12 — 8	1 + 5	7 — 3	9 — 2
11 — 7	7 + 2	5	FIVE
12 — 7	16 — 9	11 — 6	17 — 9

Addition and Subtraction Facts Dominoes
(Page 3 of six pages)
Back, laminate, and cut along dotted lines.

1 + 3	5 − 2	3	THREE
4 + 4	11 − 9	3 + 4	2 + 0
2 + 3	3 − 1	2 + 2	1 + 1
TWO	2	10 − 1	3 − 2
EIGHT	8	16 − 7	0 − 7

ADDITION AND SUBTRACTION FACTS

Addition and Subtraction Facts Dominoes
(Page 4 of six pages)
Back, laminate, and cut along dotted lines.

ADDITION AND SUBTRACTION FACTS

$1 + 4$	$1 + 2$ / $1 + 1$	$15 - 6$	$9 - 1$
$4 + 5$	$8 - 6$	$15 - 7$	$8 - 1$
$3 + 3$	$7 - 6$	$16 - 8$	$9 - 3$
$2 + 1$	$10 - 8$	$18 - 9$	$8 - 3$
$14 - 6$	$9 - 8$ / $9 - 9$	$15 - 9$	$12 - 7$

Addition and Subtraction Facts Dominoes
(Page 5 of six pages)
Back, laminate, and cut along dotted lines.

ADDITION AND SUBTRACTION FACTS

$3 + 5$	$8 - 5$	$7 - 6$	$7 - 5$
$4 + 1$	$13 - 9$	$7 - 7$	$13 - 5$
NINE	$\underline{9}$	$9 - 9$	$13 - 8$
$13 - 6$	$6 - 5$	$13 - 7$	$0 + 1$
$15 - 8$	$12 - 6$	$\underline{9}$	SIX

Addition and Subtraction Facts Dominoes
(Page 6 of six pages)
Back, laminate, and cut along dotted lines.

ADDITION AND SUBTRACTION FACTS

7	SEVEN	17 − 8	8 − 2
14 − 9	8 − 7	1 + 7	9 − 5
6 − 4	6 − 6		

Name...

Place one letter in front of each word to make a new word. Then try to place two letters in front of each word to make a new word. The first word has been done for you.

1. It - fit - slit

2. Are

3. Row

4. Rain

5. An

6. Ease

7. His

8. Eight

9. In

10. And

11. Ore

12. Art

13. One

14. At

15. Hose

MATH / LANGUAGE ARTS

Name..

Remove one letter from each word and still leave a word. Then try to remove two letters from each word and still leave a word. The first word has been done for you.

1. Four - our - or

2. Climb

3. Tried

4. Plant

5. Teen

6. Able

7. Scale

8. Cute

9. Cord

10. Rain

Multiplication Facts Table

X	0	1	2	3	4	5	6	7	8	9
0	0	0	0	0	0	0	0	0	0	0
1	0	1	2	3	4	5	6	7	8	9
2	0	2	4	6	8	10	12	14	16	18
3	0	3	6	9	12	15	18	21	24	27
4	0	4	8	12	16	20	24	28	32	36
5	0	5	10	15	20	25	30	35	40	45
6	0	6	12	18	24	30	36	42	48	54
7	0	7	14	21	28	35	42	49	56	63
8	0	8	16	24	32	40	48	56	64	72
9	0	9	18	27	36	45	54	63	72	81

Name...

Fill in the leaves.

65

28

110

18

40

6

15

30

4

10

20

2

5

10

176

Leaf Patterns
(Blank)

ame ....

Fill in the leaves.

MULTIPLICATION FACTS

1	2	3	4	5	6	7	8	9	10
11	12	13	14	15	16	17	18	19	20
21	22	23	24	25	26	27	28	29	30
31	32	33	34	35	36	37	38	39	40
41	42	43	44	45	46	47	48	49	50
51	52	53	54	55	56	57	58	59	60
61	62	63	64	65	66	67	68	69	70
71	72	73	74	75	76	77	78	79	80
81	82	83	84	85	86	87	88	89	90
91	92	93	94	95	96	97	98	99	100

Name. . . . .

Draw a line to show where each child lives.

Take the Children Home
(Blank)

Name.................................

Draw a line to show where each child lives.

MATH GAMES & ACTIVITIES. COPYRIGHT © 1984

Witch
(Multiplication facts)

Name ....

Color the picture. Twelve = black, 16 = brown, 18 = yellow, and 24 = red.

9 × 2

2 × 9

6 × 2

2 × 6

4 × 3

3 × 6

3 × 4

2 × 8

4 × 4

8 × 2

3 × 8

4 × 6

Witch
(Blank)

Name ....

Color the picture. Twelve = black, 16 = brown, 18 = yellow, and 24 = red.

Crossnumber Puzzles
(Multiplication facts)

Name...

ACROSS   DOWN

1. 9 × 9   5. 9 × 3
2. 8 × 8   6. 6 × 9
3. 3 × 0
4. 6 × 7
6. 8 × 7
7. 1 × 7
8. 2 × 2
9. 3 × 3
10. 2 × 4

ACROSS   DOWN

1. 2 × 6   1. 3 × 6
3. 6 × 6   2. 3 × 7
5. 9 × 9   3. 5 × 7
6. 9 × 6   4. 8 × 8
7. 7 × 3   7. 5 × 5
9. 4 × 8   8. 2 × 7
11. 6 × 9   9. 8 × 4
12. 4 × 7   10. 7 × 4

Crossnumber Puzzle
(Multiplication facts)

Name...

ACROSS

1. 3×8
3. 3×2
4. 8×4
5. 4×7
7. 8×7
8. 5×5
10. 3×6
12. 6×7
14. 3×5
16. 6×2
18. 2×4
19. 8×8
21. 9×8
23. 4×8
25. 9×6
26. 5×9
27. 1×3
28. 3×3

DOWN

2. 7×6
4. 6×6
6. 9×9
9. 6×9
11. 5×7
13. 7×3
15. 4×4
17. 9×3
20. 9×5
22. 6×4
24. 4×6

184

Man Gameboard
Back and laminate.
(Do not color the man.)

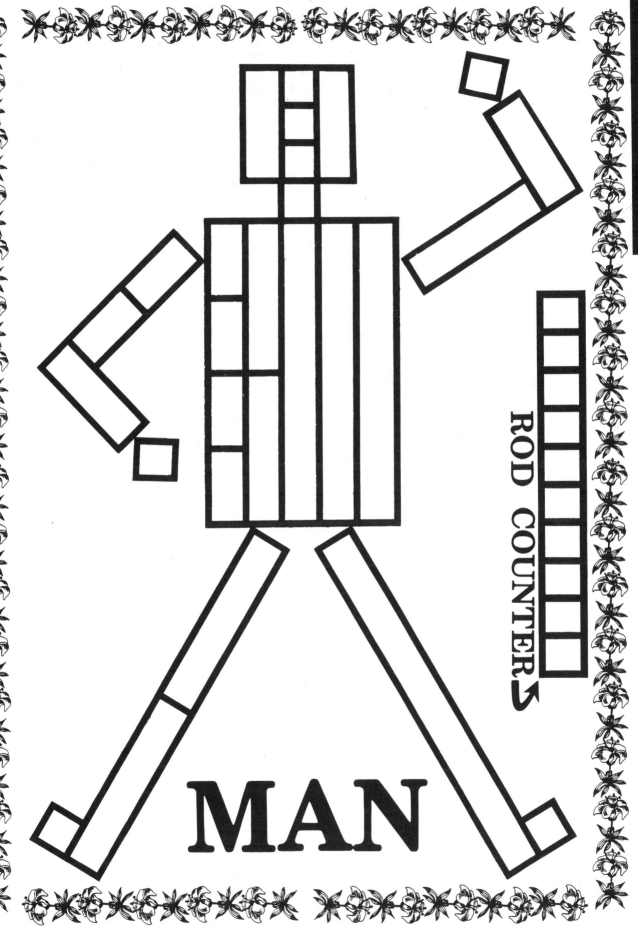

ROD COUNTERS

MAN

House Gameboard
Back and laminate.
(Do not color the house.)

HOUSE

ROD COUNTERS

Tree Gameboard
Back and laminate.
(Do **not** color the tree.)

TREE

ROD COUNTER

Caboose Gameboard
Back and laminate.
(Do not color the caboose.)

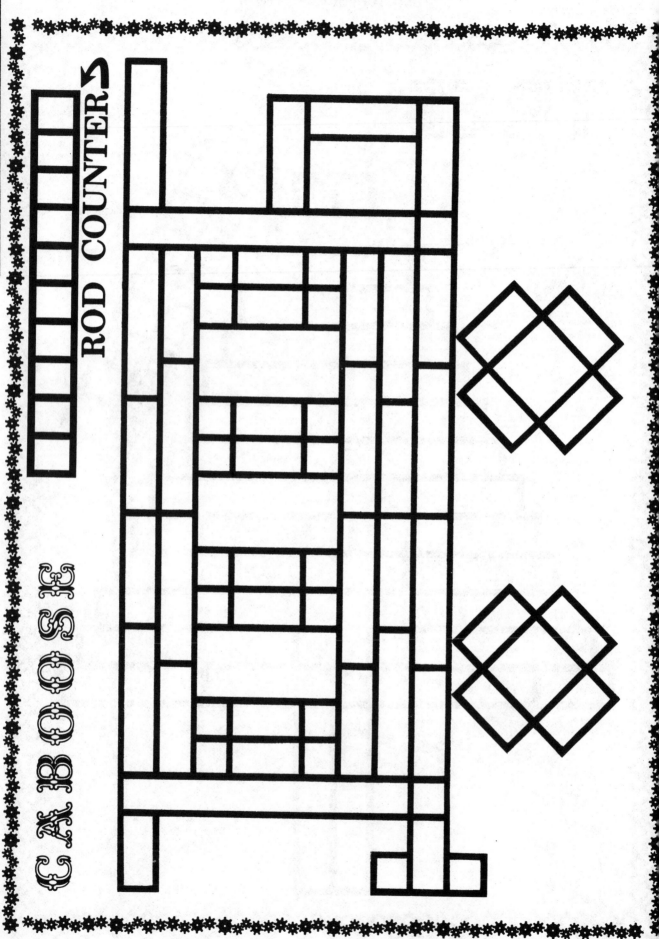

ROD COUNTERS

CABOOSE

Ring Around the Rosy Gameboard
(Multiplication facts)
Color, back, and laminate.

Ring Around the Rosy Gameboard
(Blank)

START

RING AROUND

THE ROSY

Appllingo Card
Card 1
Color, back, and laminate.
Do **not** color the apples.

APPLINGO

35

72

18

42

28

Card 1

Applingo Card
Card 2
Color, back, and laminate.
Do **not** color the apples.

Applingo Card
Card 3
Color, back, and laminate.
Do **not** color the apples.

Applingo Card
Card 4
Color, back, and laminate.
Do **not** color the apples.

APPLINGO

12

21

36

45

56

Card 4

Applingo Card
(Blank)

APPLINGO

Card

Applingo Cutouts
Page 1 of two pages
Color, back, and laminate.

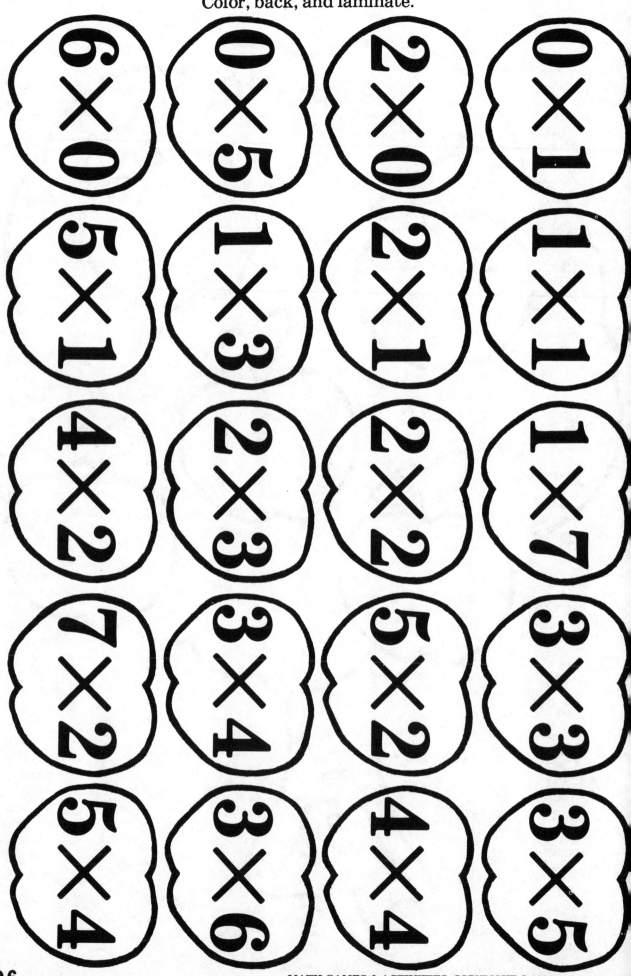

Applingo Cutouts
Page 2 of two pages
Color, back, and laminate.

MULTIPLICATION FACTS

Applingo Cutouts
(Blank)

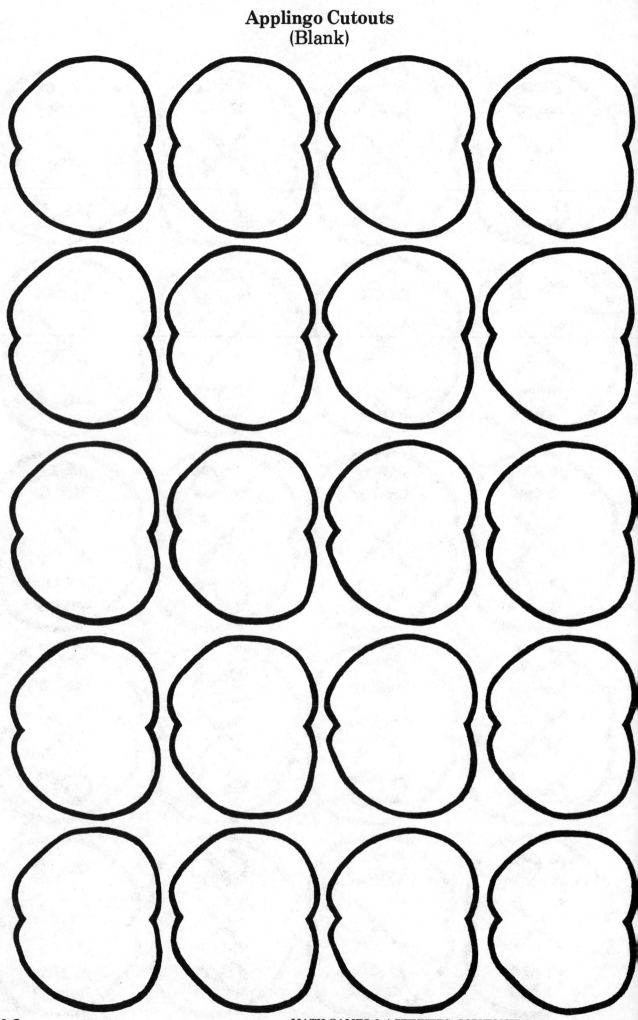

Multiplication Facts Rummy Cards
(Page 1 of seven pages)
Back, laminate, and cut along solid lines.

MULTIPLICATION FACTS

3×2

3 × 2

2×3

2 × 3

6

6

4×3

4 × 3

3×4

3 × 4

12

12

Multiplication Facts Rummy Cards
(Page 2 of seven pages)
Back, laminate, and cut along solid lines.

MULTIPLICATION FACTS

5 × 4

5×4

$ $ $ $ $
$ $ $ $ $
$ $ $ $ $
$ $ $ $ $

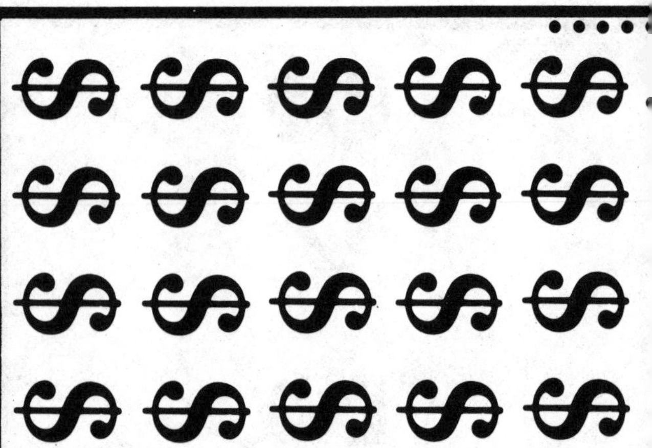

20

20

4 × 5

4×5

6 × 3

6×3

★ ★ ★ ★ ★ ★
★ ★ ★ ★ ★ ★
★ ★ ★ ★ ★ ★

18

18

3 × 6

3×6

Multiplication Facts Rummy Cards
(Page 3 of seven pages)
Back, laminate, and cut along solid lines.

6×4

6 × 4

4×6

4 × 6

24

24

7×3

7 × 3

3×7

3 × 7

21

21

Multiplication Facts Rummy Cards
(Page 4 of seven pages)
Back, laminate, and cut along solid lines.

7 × 4

7 × 4

28

4 × 7

4 × 7

7 × 6

7 × 6

42

6 × 7

6 × 7

Multiplication Facts Rummy Cards
(Page 5 of seven pages)
Back, laminate, and cut along solid lines.

MULTIPLICATION FACTS

8×4

8 × 4

4×8

4 × 8

32

32

8×6

8 × 6

6×8

6 × 8

48

48

Multiplication Facts Rummy Cards
(Page 6 of seven pages)
Back, laminate, and cut along solid lines.

MULTIPLICATION FACTS

8 × 7

8 × 7

56

56

7 × 8

8 × 8

9 × 8

9 × 8

8 × 9

72

72

8 × 9

Multiplication Facts Rummy Cards
(Page 7 of seven pages)
Back, laminate, and cut along solid lines.

MULTIPLICATION FACTS

9×7

9 × 7

7×9

7 × 9

63

63

Name...

DIVISION FACTS

Color the picture. One = yellow, 2 = red, 3 = green, 4 = blue, 5 = brown, and 6 = pink.

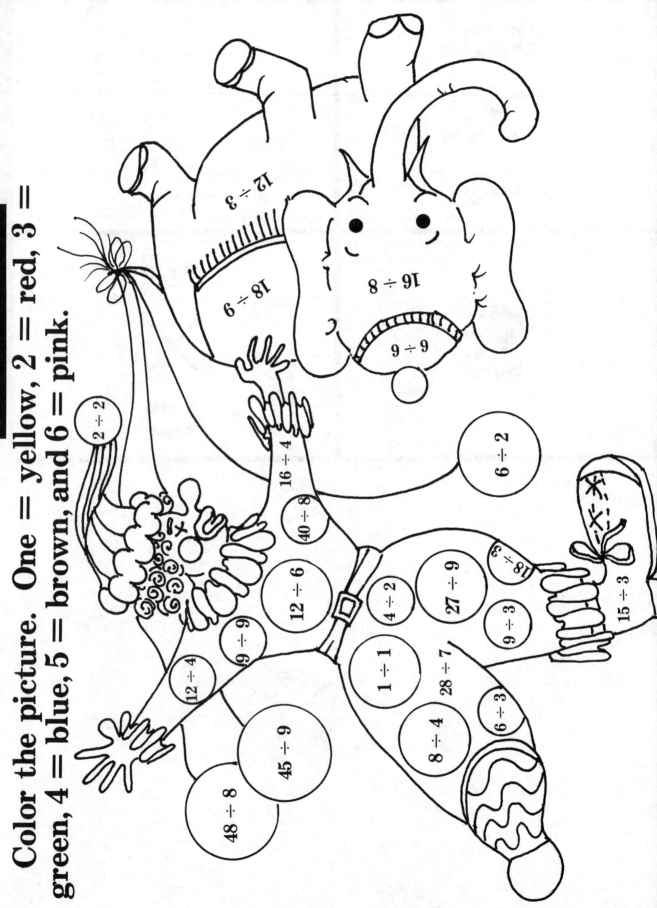

206 MATH GAMES & ACTIVITIES. COPYRIGHT © 1984

Name...

Color the picture. One = yellow, 2 = red, 3 = green, 4 = blue, 5 = brown, and 6 = pink.

DIVISION FACTS

Ski Slope Gameboard
(Division facts)
Top
Color, back, and laminate.

START

Hit fresh pow- der. Ski two more spaces.

SKI SLOPE

Ran into a tree. Sit out one turn.

DIVISION FACTS

Ski Slope Gameboard
(Division facts)
Bottom
Color, back, and laminate.

Lost a pole. Sit out one turn.

Great paralleling. Ski two more spaces.

LIFT

Division Cards
(Page 1 of two pages)
Back, laminate, and cut along solid lines.

DIVISION FACTS

DIVISION CARD $2\overline{)5}$	DIVISION CARD $2\overline{)14}$	DIVISION CARD $2\overline{)17}$	DIVISION CARD $3\overline{)8}$
DIVISION CARD $3\overline{)16}$	DIVISION CARD $3\overline{)27}$	DIVISION CARD $4\overline{)10}$	DIVISION CARD $4\overline{)21}$
DIVISION CARD $4\overline{)28}$	DIVISION CARD $4\overline{)35}$	DIVISION CARD $5\overline{)16}$	DIVISION CARD $5\overline{)25}$
DIVISION CARD $5\overline{)32}$	DIVISION CARD $5\overline{)44}$	DIVISION CARD $6\overline{)18}$	DIVISION CARD $6\overline{)27}$
DIVISION CARD $6\overline{)40}$	DIVISION CARD $6\overline{)50}$	DIVISION CARD $7\overline{)11}$	DIVISION CARD $7\overline{)26}$

Division Cards
(Page 2 of two pages)
Back, laminate, and cut along solid lines.

DIVISION CARD	DIVISION CARD	DIVISION CARD	DIVISION CARD
7)42	7)53	8)20	8)25
DIVISION CARD	DIVISION CARD	DIVISION CARD	DIVISION CARD
8)50	8)72	9)23	9)46
DIVISION CARD	DIVISION CARD	DIVISION CARD	DIVISION CARD
9)63	9)76	10)33	10)42
DIVISION CARD	DIVISION CARD	DIVISION CARD	DIVISION CARD
10)64	10)90	11)59	11)77
DIVISION CARD	DIVISION CARD	DIVISION CARD	DIVISION CARD
11)90	12)48	12)63	12)88

DIVISION FACTS

Name . . . .

	48

8	8
8	8	8	8

Now you balance the rest.

6	6	6	6
6	6	6	6

56	

	7	7

54	

	9	9

32	

8			8

	5	5
5	5	5	5

	4	4
	4	4

Teeter Totter
(Blank)

ame . . . .

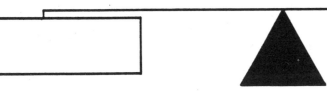

Now you balance the rest.

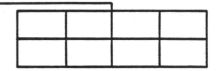

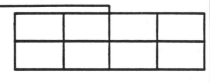

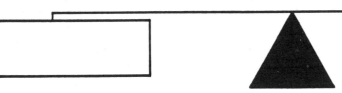

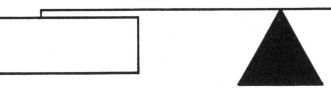

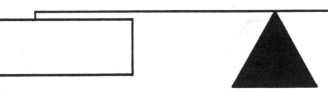

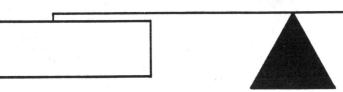

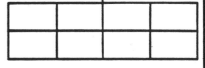

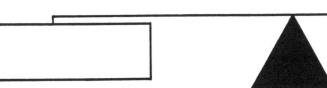

MULTIPLICATION AND DIVISION FACTS

Crossnumber Puzzle
(Multiplication and division facts)

Name..

ACROSS

1. 8×5
2. $54 \div 9$
3. 3×8
5. $24 \div 6$
6. 9×6
7. 6×8
8. $45 \div 9$
9. $63 \div 9$
10. $21 \div 7$
11. 9×7
13. 5×9
14. $48 \div 8$
15. 7×8
17. $56 \div 8$
18. $72 \div 8$
19. 5×6
20. 9×5
21. 9×9

DOWN

1. 6×7
2. 8×8
3. 8×3
4. 8×6
5. 7×7
8. 8×7
9. 9×8
10. 5×7
12. 7×5
14. 7×9
16. 8×8

214

Hi! My name is Ima Blank-page. Turn the page and BEHOLD!

Stepping Stones Gameboard
(Multiplication and division facts)
Left side
Color, back, and laminate.

Stepping Stones Gameboard
(Multiplication and division facts)
Right side
Color, back, and laminate.

START

2×2

2×3

1×1

$6 \div 2$

$6 \div 3$

$10 \div 2$

$16 \div 4$

1×2

$35 \div 7$

6×1

$21 \div 7$

$8 \div 8$

STONES

STEPPING

FINISH

Stepping Stones Gameboard
(Blank)
Right side

START

STONES

MIXED FACTS

Name . . . .

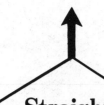

Straight

$4 \div 1 + 6 =$

$14 \div 2 + 2 =$

$12 \div 3 + 4 =$

$32 \div 4 - 1 =$

$45 \div 5 - 3 =$

$36 \div 6 - 1 =$

$42 \div 7 - 2 =$

$24 \div 8 =$

$18 \div 9 =$

$5 \div 5 =$

Blast off!

Crazy

$36 \div 4 + 1 =$

$16 \div 2 - 2 =$

$40 \div 5 - 4 =$

$6 \div 2 + 2 =$

$56 \div 7 + 1 =$

$16 \div 8 + 1 =$

$49 \div 7 + 1 =$

$14 \div 7 =$

$63 \div 9 =$

$9 \div 9 =$

Blast off!

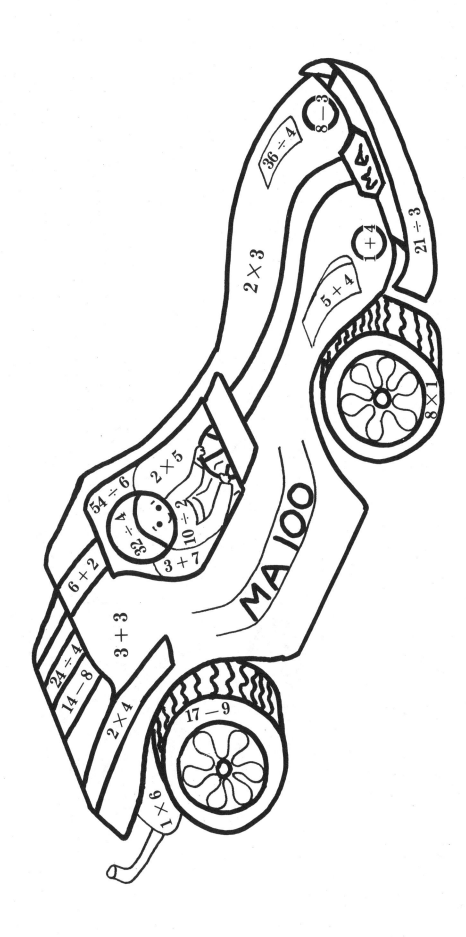

Racer
(Mixed facts)

Name...............

Color the picture. Five = yellow, 6 = blue, 7 = silver, 8 = red, 9 = black, and 10 = brown.

Racer
(Blank)

Name.................................

Color the picture. Five = yellow, 6 = blue, 7 = silver, 8 = red, 9 = black, and 10 = brown.

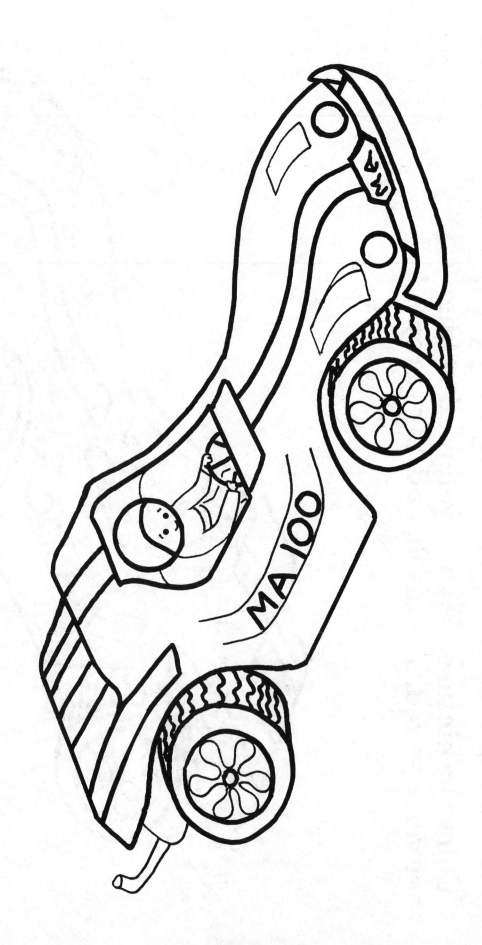

222

Moon Walk Gameboard
Color, back, and laminate.

MOON

MOON WALK

LUNAR STATION

18	24	32	45
2	3	8	9

STARSHIP

12	21	30	42
1	5	6	10

EARTH

223

Trizo Gameboard
Color, back, and laminate.

T R I Z O

1	2	3	4	5	6	7	8
9	10	11	12	13	14	15	16
17	18	19	20	21	22	23	24
25	26	27	28	29	30	FREE	32
33	34	35	36	37	38	39	40
41	42	FREE	45	48	50	54	55
60	64	66	72	75	FREE	90	96
100	108	120	125	144	150	180	216

Improve Your Aim
(Addition)

Name..

Score exactly 100 in the least number of shots.

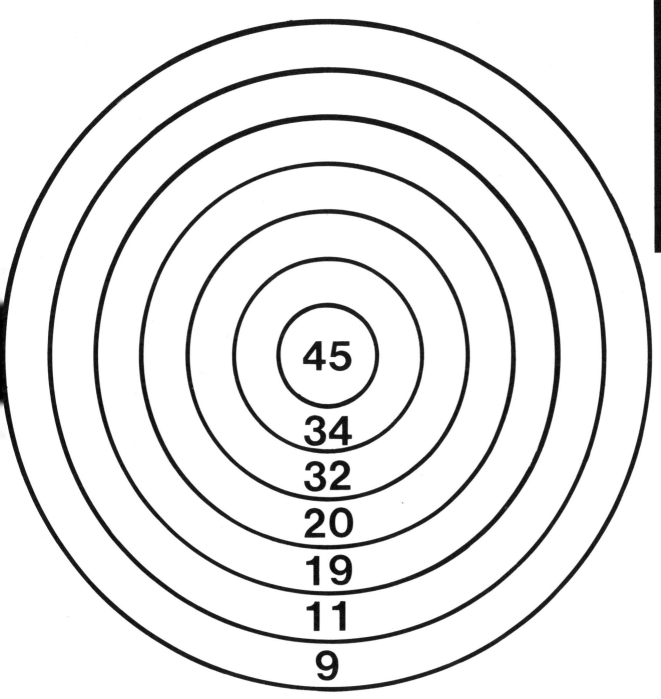

45
34
32
20
19
11
9

Name...

Score exactly _____ in the least number of shots.

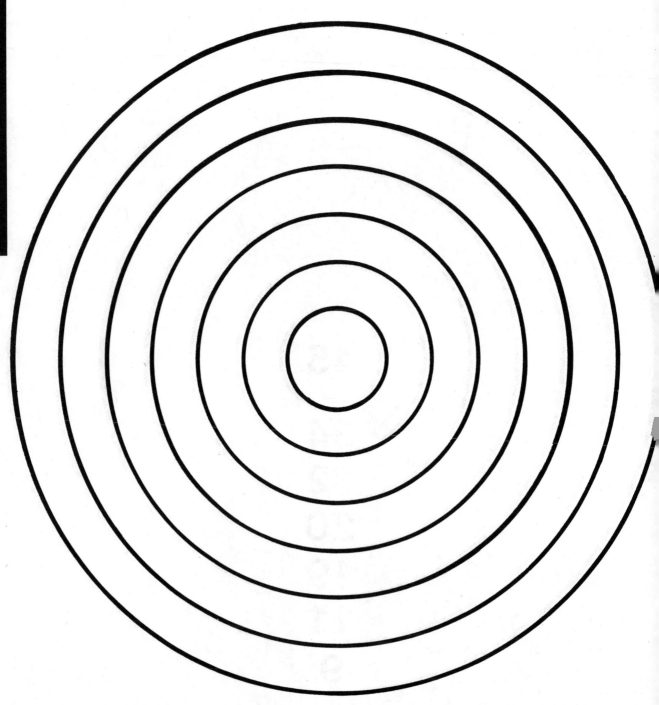

Fly, Fly Away
(Subtraction)

Name...

Lighten the balloon to exactly   1250   by dropping one sandbag.

1547

297   147   397   247

Name...

Lighten the balloon to exactly   by dropping one sandbag.

Subtraction Magic

Name............................

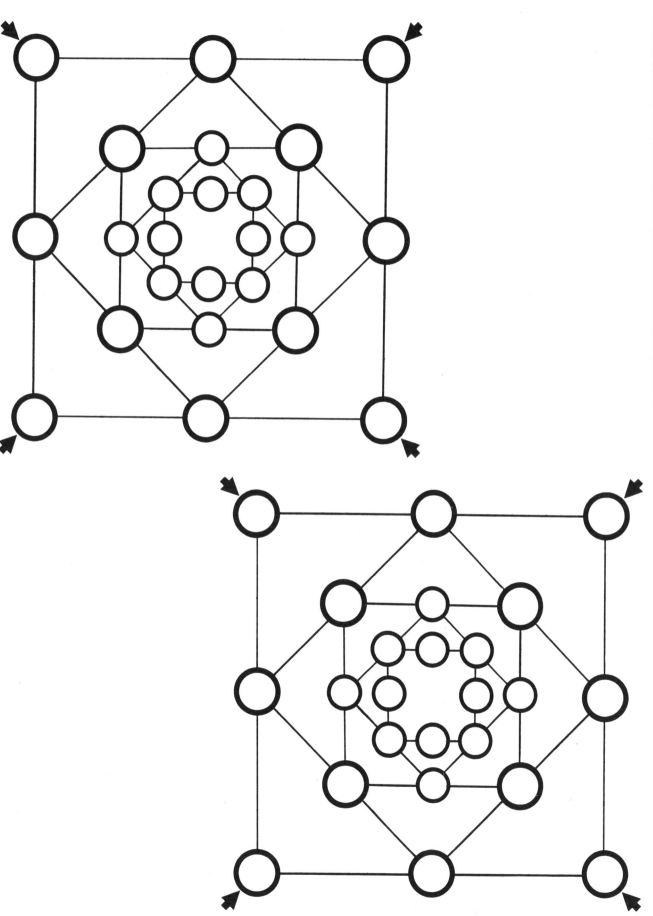

Name.............................

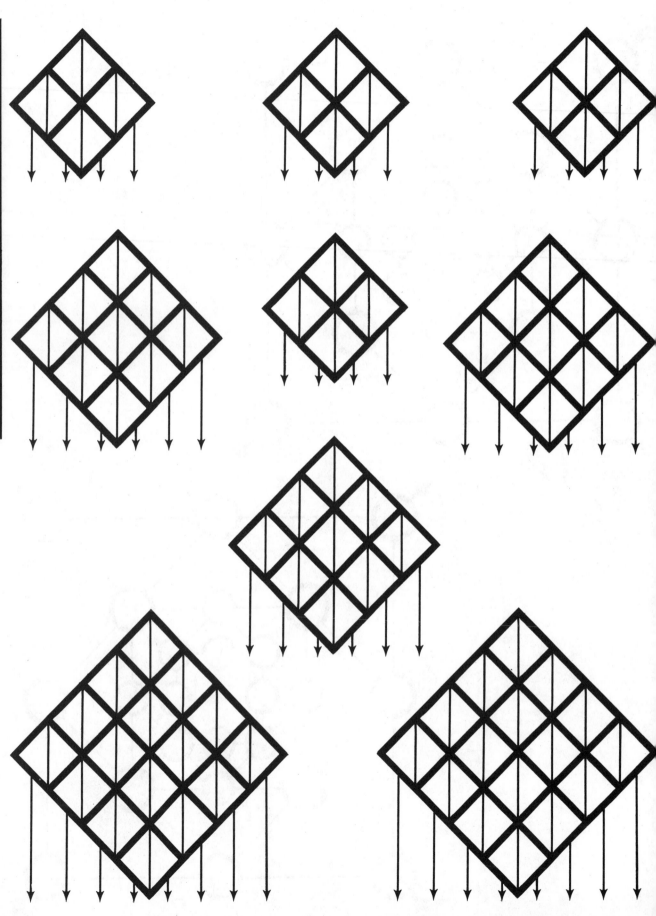

WHOLE NUMBER ARITHMETIC

Division Worksheet

WHOLE NUMBER ARITHMETIC

100	100	100	100	100	100	100	100	100	100

10 10	10 10	10 10	10 10	10 10	10 10	10 10	10 10	10 10	10 10
10 10	10 10	10 10	10 10	10 10	10 10	10 10	10 10	10 10	10 10
10 10	10 10	10 10	10 10	10 10	10 10	10 10	10 10	10 10	10 10
10 10	10 10	10 10	10 10	10 10	10 10	10 10	10 10	10 10	10 10
10 10	10 10	10 10	10 10	10 10	10 10	10 10	10 10	10 10	10 10
10	10	10	10	10	10	10	10	10	10

1	1	1	1	1	1	1	1	1	1
1	1	1	1	1	1	1	1	1	1
1	1	1	1	1	1	1	1	1	1
1	1	1	1	1	1	1	1	1	1
1	1	1	1	1	1	1	1	1	1
1	1	1	1	1	1	1	1	1	1
1	1	1	1	1	1	1	1	1	1
1	1	1	1	1	1	1	1	1	1
1	1	1	1	1	1	1	1	1	1
1	1	1	1	1	1	1	1	1	1

R

Speedy Operator
(Blank)

SPEEDY OPERATOR

WINNER'S CIRCLE

STARTER BOX

SPIDER AND FLY

WHOLE NUMBER ARITHMETIC

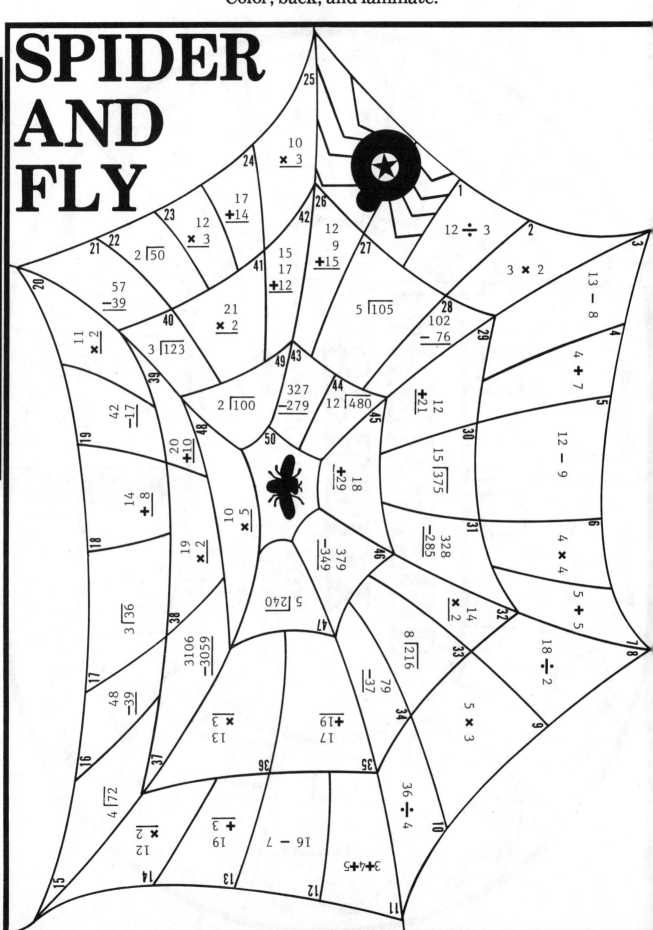

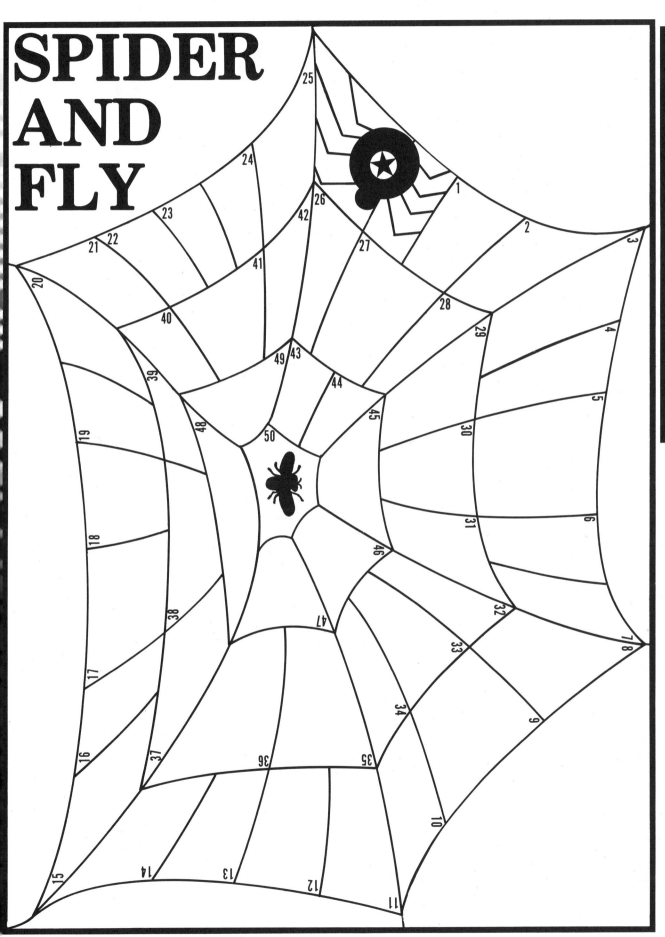

SPIDER
AND
FLY

Doll Gameboard
Color, back, and laminate.

DOLL

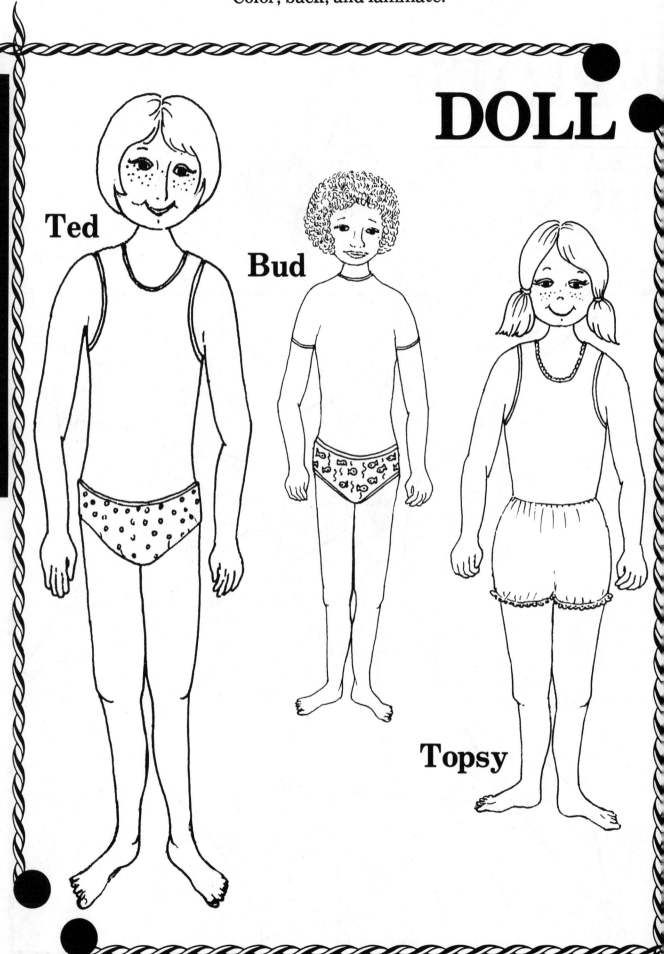

Ted

Bud

Topsy

Doll Cutouts
(Clothes for Ted)
Color, back, and laminate.

Doll Cutouts
(Clothes for Bud)
Color, back, and laminate.

Doll Cutouts
(Clothes for Topsy)
Color, back, and laminate.

Doll Cards

Back, laminate, and cut along solid lines.

DOLL One of your dolls is having a birthday. Take two items of clothing for whichever doll you choose.	DOLL It's Christmas! Take any three items of clothing for your dolls.
DOLL One of your dolls forgot to weat a bib while eating spaghetti. Return a shirt or a blouse.	DOLL One of your dolls spilled a drink on its lap. Return a skirt or a pair of pants.
DOLL It's a sunny day. Take a hat for one of your dolls.	DOLL No forgetting to bring your dolls in out of the rain. Miss a turn.
DOLL Sorry, but you can't find your dolls. Miss a turn.	DOLL No forgetting the names of your dolls. Miss a turn.

Hi! My name is Ima Blank-page. Turn the page and BEHOLD!

Doll House Gameboard
Left side
Color, back, and laminate.

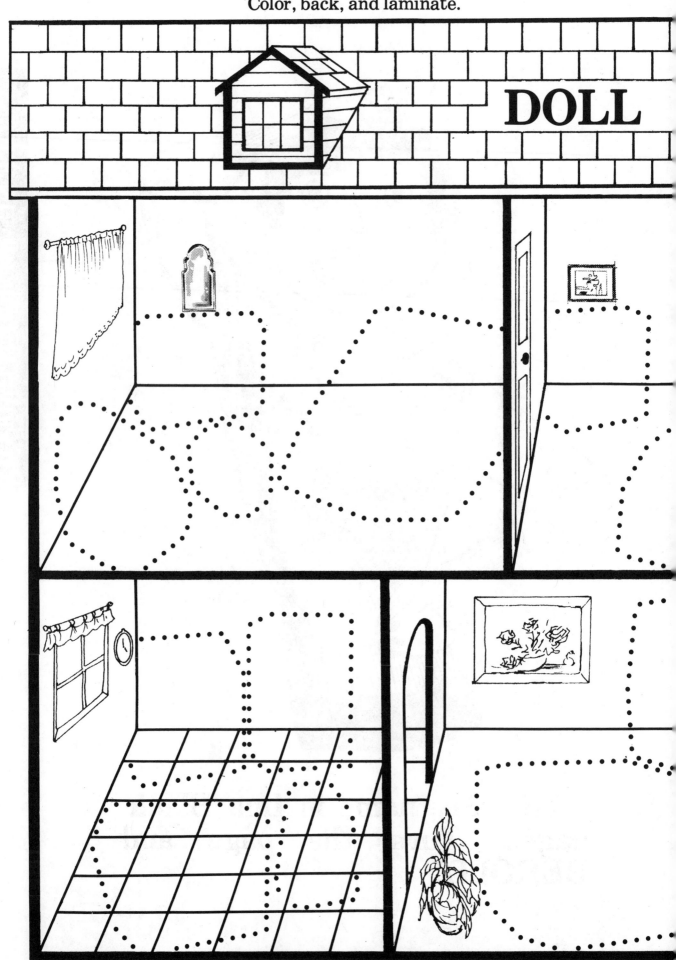

DOLL

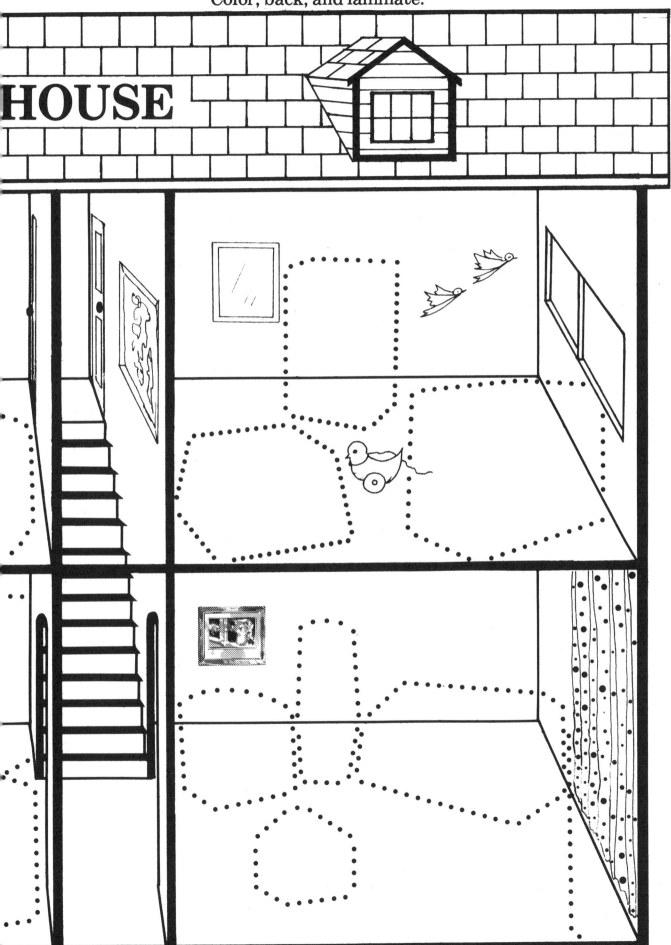

HOUSE

Doll House Cutouts
(Kitchen, den, and master bedroom furniture)
Color, back, and laminate.

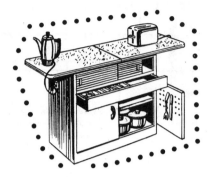

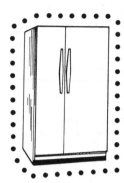

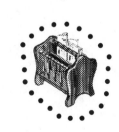

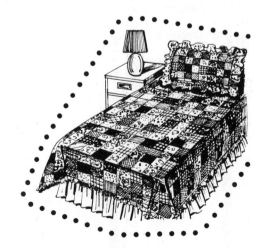

Doll House Cutouts
(Dining Room, living room, and child's bedroom furniture)
Color, back, and laminate.

WHOLE NUMBER ARITHMETIC

DOLL HOUSE It's dinner time! Take the table and chairs.	DOLL HOUSE Beds are not for jumping on. Return the bed.
DOLL HOUSE No putting your shoes on the couch. Return the couch.	DOLL HOUSE No leaving the refrigerator door open. Return the refrigerator.
DOLL HOUSE Just in case you want to read, take the lamp.	DOLL HOUSE Sorry, but you locked yourself out of the house. Miss a turn.
DOLL HOUSE No leaving your clothes on the floor. Miss a turn.	DOLL HOUSE No forgetting to turn the lights off. Miss a turn.

World Series Gameboard
Color, back, and laminate.

WHOLE NUMBER ARITHMETIC

WORLD SERIES Base hit! Run one base.	**WORLD SERIES** Ball four! Walk one base.
WORLD SERIES Double! Run two bases.	**WORLD SERIES** Home run! Jog four bases.
WORLD SERIES Pitcher off guard! Steal one base.	**WORLD SERIES** Strike three! Play goes to opponent.
WORLD SERIES Pop fly! Play goes to opponent.	**WORLD SERIES** Thrown out at first base! Play goes to opponent.

Super Bowl Gameboard
Right side **and** left side
Join at 50-yard line.
Color, back, and laminate.

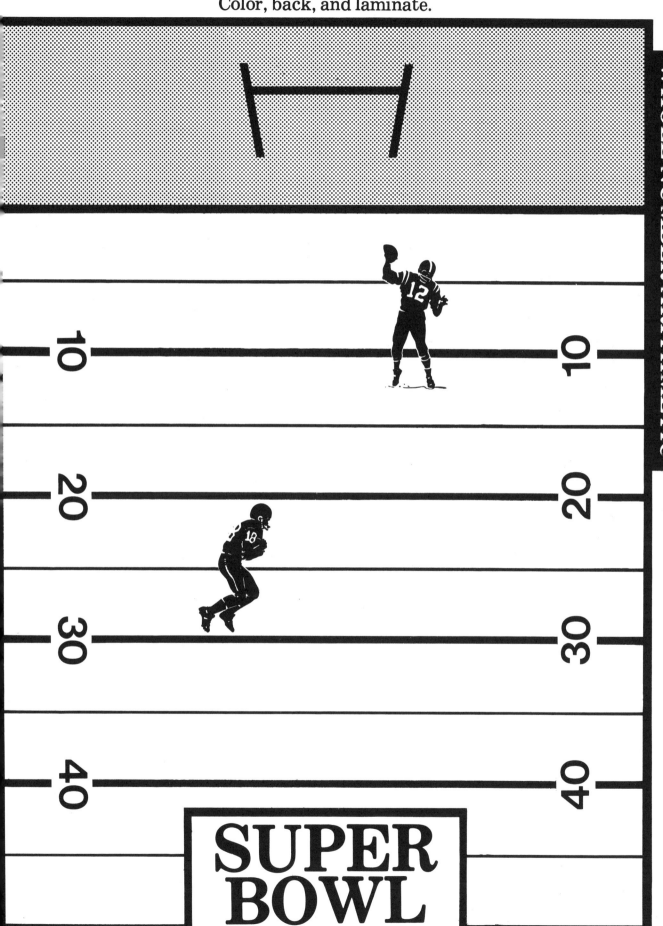

WHOLE NUMBER ARITHMETIC

10   10

20   20

30   30

40   40

SUPER BOWL

WHOLE NUMBER ARITHMETIC

SUPER BOWL

Outstanding run! Go forward 45 yards.

SUPER BOWL

Off sides! Go back 5 yards.

SUPER BOWL

Broken play! Go back 10 yards.

SUPER BOWL

Holding! Go back 15 yards.

SUPER BOWL

Completed pass! Go forward 30 yards.

SUPER BOWL

Quarter back sacked on fourth down! Play goes to opponent.

SUPER BOWL

Intercepted pass! Play goes to opponent.

SUPER BOWL

Fumble! Play goes to opponent.

ASMD Cards
(Page 1 of four pages)
Back, laminate, and cut along solid lines.

1

ASMD

Do just one.

$3 + 3 = ?$
$8 - 1 = ?$
$5 \times 2 = ?$
$16 \div 4 = ?$

2

ASMD

Do just one.

$4 + 1 = ?$
$6 - 3 = ?$
$8 \times 1 = ?$
$4 \div 2 = ?$

3

ASMD

Do just one.

$2 + 6 = ?$
$9 - 7 = ?$
$7 \times 0 = ?$
$9 \div 3 = ?$

4

ASMD

Do just one.

$7 + 9 = ?$
$7 - 4 = ?$
$3 \times 6 = ?$
$45 \div 9 = ?$

5

ASMD

Do just one.

$3 + 4 + 5 = ?$
$17 - 8 = ?$
$6 \times 7 = ?$
$81 \div 9 = ?$

6

ASMD

Do just one.

$9 + 6 + 1 = ?$
$12 - 5 = ?$
$9 \times 8 = ?$
$64 \div 8 = ?$

7

ASMD

Do just one.

$2 + 5 + 1 = ?$
$13 - 6 = ?$
$4 \times 7 = ?$
$35 \div 5 = ?$

8

ASMD

Do just one.

$6 + 2 + 5 = ?$
$14 - 9 = ?$
$8 \times 5 = ?$
$56 \div 7 = ?$

WHOLE NUMBER ARITHMETIC

ASMD Cards
(Page 2 of four pages)
Back, laminate, and cut along solid lines.

WHOLE NUMBER ARITHMETIC

9

ASMD
Do just one.

$$263 \atop +415$$ $$573 \atop -152$$

$$23 \atop \times\ 2$$ $3\overline{)65}$

10

ASMD
Do just one.

$$362 \atop +\ 15$$ $$567 \atop -\ 45$$

$$97 \atop \times\ 1$$ $4\overline{)74}$

11

ASMD
Do just one.

$$54 \atop +43$$ $$75 \atop -23$$

$$312 \atop \times\ 3$$ $6\overline{)84}$

12

ASMD
Do just one.

$$2064 \atop +3512$$ $$5206 \atop -1102$$

$$1212 \atop \times\ \ 4$$ $5\overline{)90}$

13

ASMD
Do just one.

$$38 \atop +25$$ $$53 \atop -24$$

$$25 \atop \times\ 3$$ $2\overline{)532}$

14

ASMD
Do just one.

$$44 \atop +\ 6$$ $$62 \atop -\ 7$$

$$36 \atop \times\ 2$$ $5\overline{)608}$

15

ASMD
Do just one.

$$37 \atop +59$$ $$56 \atop -18$$

$$124 \atop \times\ \ 4$$ $7\overline{)784}$

16

ASMD
Do just one.

$$135 \atop +426$$ $$295 \atop -168$$

$$1116 \atop \times\ \ 5$$ $8\overline{)915}$

ASMD Cards
(Page 3 of four pages)
Back, laminate, and cut along solid lines.

17

ASMD

Do just one.

$$352 + 184$$ $$347 - 165$$

$$63 \times 2$$ $$2\overline{)4628}$$

18

ASMD

Do just one.

$$295 + 31$$ $$439 - 78$$

$$53 \times 3$$ $$5\overline{)6085}$$

19

ASMD

Do just one.

$$473 + 363$$ $$615 - 242$$

$$190 \times 5$$ $$3\overline{)51642}$$

20

ASMD

Do just one.

$$4562 + 1249$$ $$3624 - 2158$$

$$2132 \times 4$$ $$4\overline{)910526}$$

21

ASMD

Do just one.

$$365 + 287$$ $$524 - 335$$

$$37 \times 4$$ $$2\overline{)416}$$

22

ASMD

Do just one.

$$234 + 596$$ $$761 - 297$$

$$26 \times 8$$ $$3\overline{)903}$$

23

ASMD

Do just one.

$$466 + 266$$ $$932 - 486$$

$$173 \times 5$$ $$8\overline{)1630}$$

24

ASMD

Do just one.

$$3442 + 1278$$ $$5611 - 1355$$

$$2069 \times 3$$ $$7\overline{)21357}$$

WHOLE NUMBER ARITHMETIC

ASMD Cards
(Page 4 of four pages)
Back, laminate, and cut along solid lines.

WHOLE NUMBER ARITHMETIC

25

ASMD
Do just one.

```
 345          5210
  26        −1436
+458
```

```
 345
×   6         24⟌336
```

26

ASMD
Do just one.

```
  59         82153
 436       −65062
+157
```

```
 827
×   5         15⟌800
```

27

ASMD
Do just one.

```
 413         32576
 255       −13688
+164
```

```
8349
×   3         31⟌6541
```

28

ASMD
Do just one.

```
 2124        621345
 3246      −242569
+1387
```

```
23157
×     9        42⟌9136
```

29

ASMD
Do just one.

```
 3258        8030
 1467      −1526
+2903
```

```
 45
×27           36⟌720
```

30

ASMD
Do just one.

```
 487         9000
 396       −1234
+721
```

```
 36
×48           88⟌935
```

31

ASMD
Do just one.

```
 41362       2400
 12503     −1632
+28356
```

```
 134
× 35          24⟌4920
```

32

ASMD
Do just one.

```
38001        50300
 2543      −  2436
  496
+ 3812
```

```
2159
×  16         15⟌6050
```

ASMD CARDS

KEY

Card	+	−	×	÷	Card	+	−	×	÷
1	6	7	10	4	17	536	182	126	2314
2	5	3	8	2	18	326	361	159	1217
3	8	2	0	3	19	836	373	950	17,214
4	16	3	18	5	20	5811	1466	8528	227,631 R2
5	12	9	42	9	21	652	189	148	208
6	16	7	72	8	22	830	464	208	301
7	8	7	28	7	23	732	446	865	203 R6
8	13	5	40	8	24	4720	4256	6207	3051
9	678	421	46	21 R2	25	829	3774	2070	14
10	377	522	97	18 R2	26	652	17,091	4135	53 R5
11	97	52	936	14	27	832	18,888	25,047	211
12	5576	4104	4848	18	28	6757	378,776	208,413	217 R22
13	63	29	75	266	29	7628	6504	1215	20
14	50	55	72	121 R3	30	1604	7766	1728	10 R55
15	96	38	496	112	31	82,221	768	4690	205
16	561	127	5580	114 R3	32	44,852	47,864	34,544	403 R5

ASMD Scorecard

Name... Date...............

WHOLE NUMBER ARITHMETIC

Card	Operation				Card	Operation			
	+	**−**	**×**	**÷**		**+**	**−**	**×**	**÷**
	Hit Miss	Hit Miss	Hit Miss	Hit Miss		Hit Miss	Hit Miss	Hit Miss	Hit Miss
1	— —	— —	— —	— —	17	— —	— —	— —	— —
2	— —	— —	— —	— —	18	— —	— —	— —	— —
3	— —	— —	— —	— —	19	— —	— —	— —	— —
4	— —	— —	— —	— —	20	— —	— —	— —	— —
5	— —	— —	— —	— —	21	— —	— —	— —	— —
6	— —	— —	— —	— —	22	— —	— —	— —	— —
7	— —	— —	— —	— —	23	— —	— —	— —	— —
8	— —	— —	— —	— —	24	— —	— —	— —	— —
9	— —	— —	— —	— —	25	— —	— —	— —	— —
10	— —	— —	— —	— —	26	— —	— —	— —	— —
11	— —	— —	— —	— —	27	— —	— —	— —	— —
12	— —	— —	— —	— —	28	— —	— —	— —	— —
13	— —	— —	— —	— —	29	— —	— —	— —	— —
14	— —	— —	— —	— —	30	— —	— —	— —	— —
15	— —	— —	— —	— —	31	— —	— —	— —	— —
16	— —	— —	— —	— —	32	— —	— —	— —	— —

256

Motley Crab Adder

UNDERSTANDING ARITHMETIC WORD PROBLEMS

The Scruffy Twin Subtractors

Sir Crab Multiplier

The Impeccable Twin Dividers

UNDERSTANDING ARITHMETIC WORD PROBLEMS

The Great Legalizer

The Magnificent Equalizer

Motley and Mates
(Miniatures)

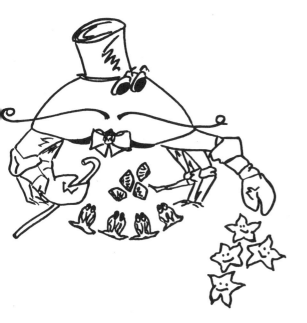

Dot Paper
(2-centimeter)

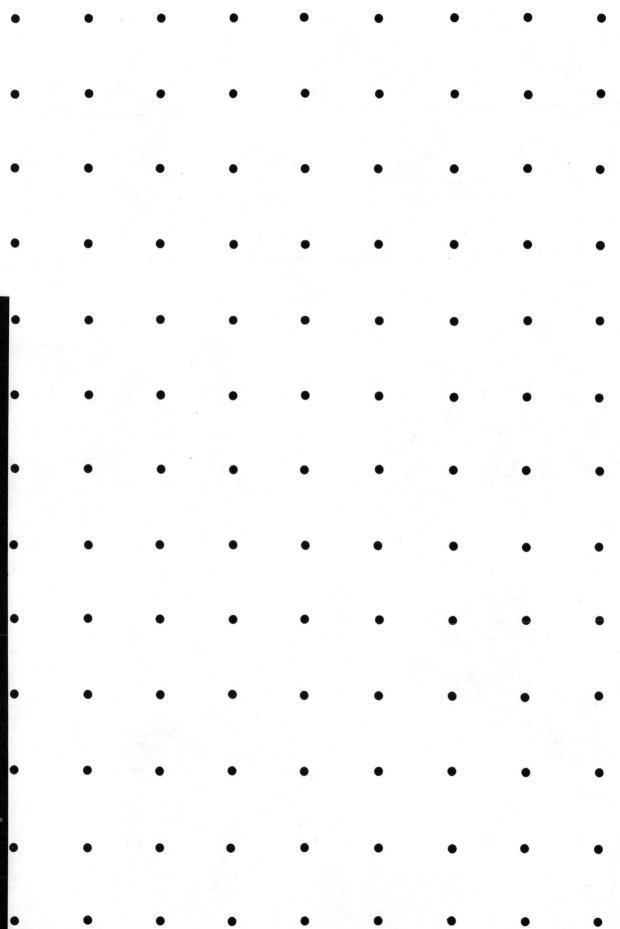

Dot Paper
(1-centimeter)

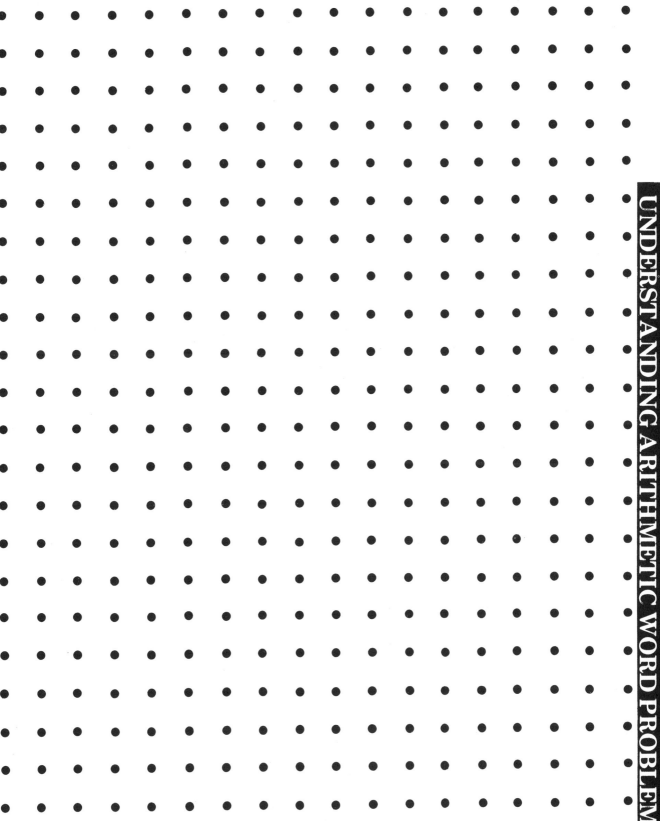

Shopping Cards
(Page 1 of three pages)
Color, back, and laminate.
Cut along solid lines.

Grapes 2¢

1¢ **Strawberries**

Cherries 3¢

Root beer 4¢

Ground beef 6¢

5¢ **Cheese**

Strip steak 7¢

8¢ **Pork roast**

UNDERSTANDING ARITHMETIC WORD PROBLEMS

Shopping Cards
(Page 2 of three pages)
Color, back, and laminate.
Cut along solid lines.

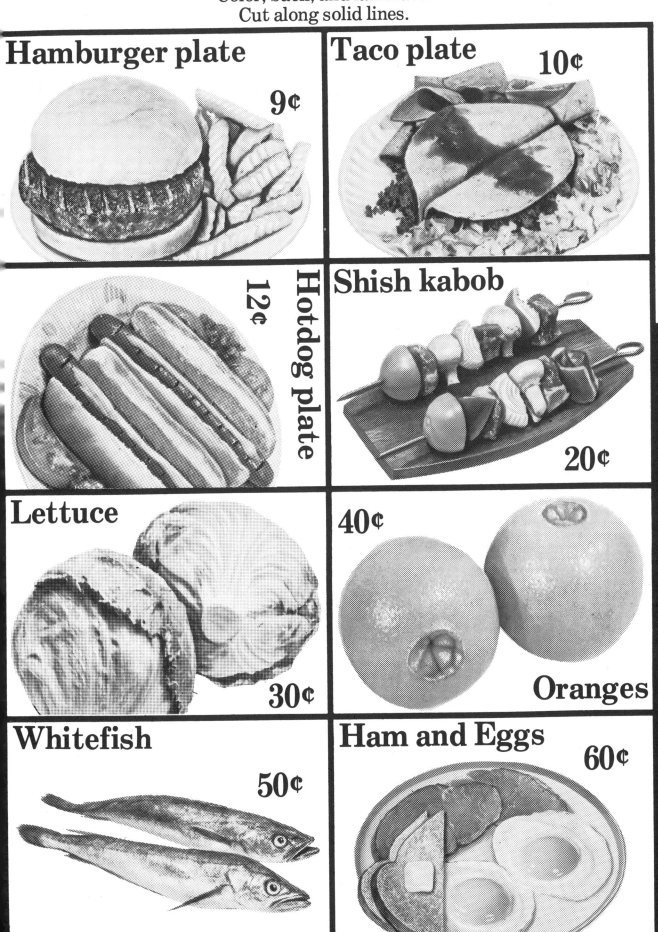

Hamburger plate

9¢

Taco plate

10¢

Hotdog plate

12¢

Shish kabob

20¢

Lettuce

30¢

40¢

Oranges

Whitefish

50¢

Ham and Eggs

60¢

UNDERSTANDING ARITHMETIC WORD PROBLEMS

Shopping cards
(Page 3 of three pages)
Color, back, and laminate.
Cut along solid lines.

Donuts

75¢

Pumpkins

90¢

Nectarines **$1.20**

$1.00

Apples

Carrots

$1.50

$1.80

Eggs

Corn **$2.40**

$3.00

Tarts

UNDERSTANDING ARITHMETIC WORD PROBLEMS

MATH GAMES & ACTIVITIES. COPYRIGHT © 1984

Fraction Rulers
(Unit, halves, fourths, fifths, and sixths)

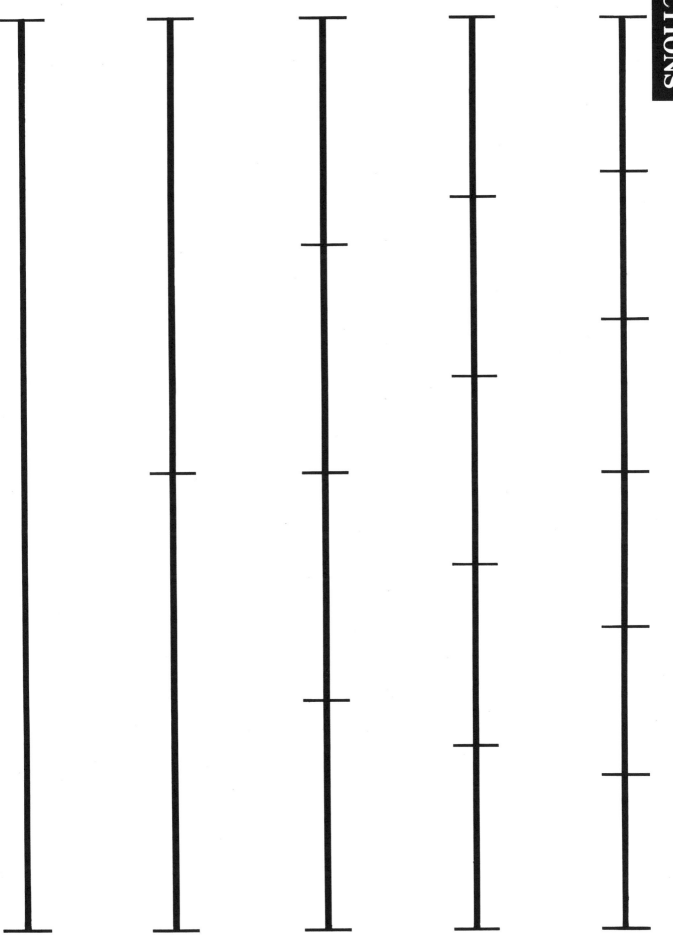

Fraction Rulers
(Unit, eighths, tenths, twelfths, and sixteenths)

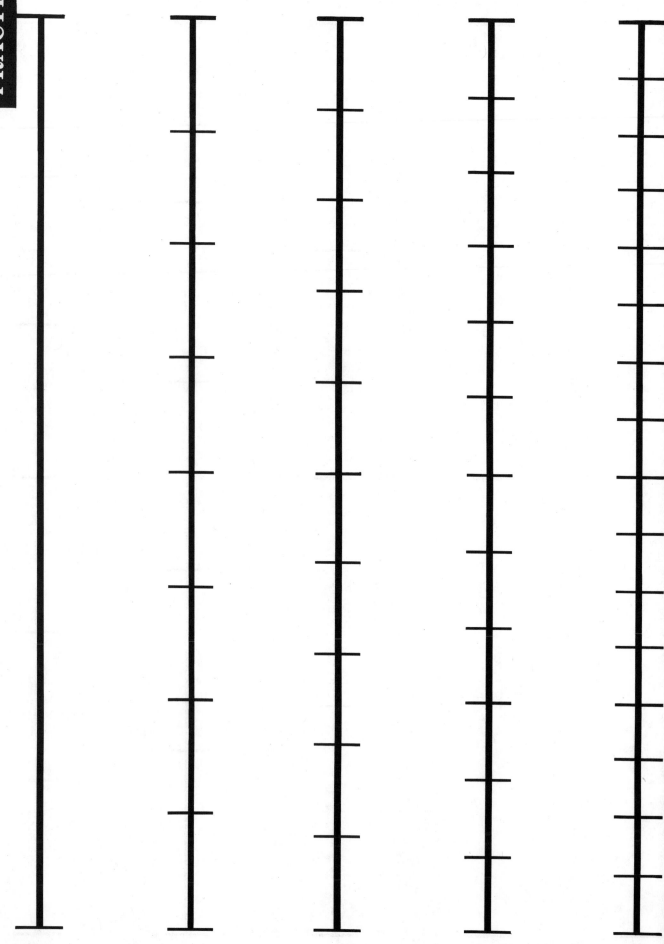

Name.....................................

MERRY MEASURING

Cut a piece of string equal to your height.

Fold it in half and try it on yourself.

What can you find that is $\frac{1}{2}$ your height?

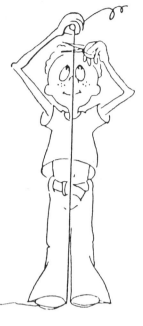

Fold it in thirds. What can you find that's $\frac{1}{3}$ your height? $\frac{1}{4}$? $\frac{1}{5}$? What else?

RECORD HERE

$\frac{1}{2}$ my height	$\frac{1}{3}$ my height	$\frac{1}{4}$ my height	$\frac{1}{5}$ my height	What else?

Fraction Dominoes
(Page 1 of three pages)
Back, laminate, and cut along dotted lines.

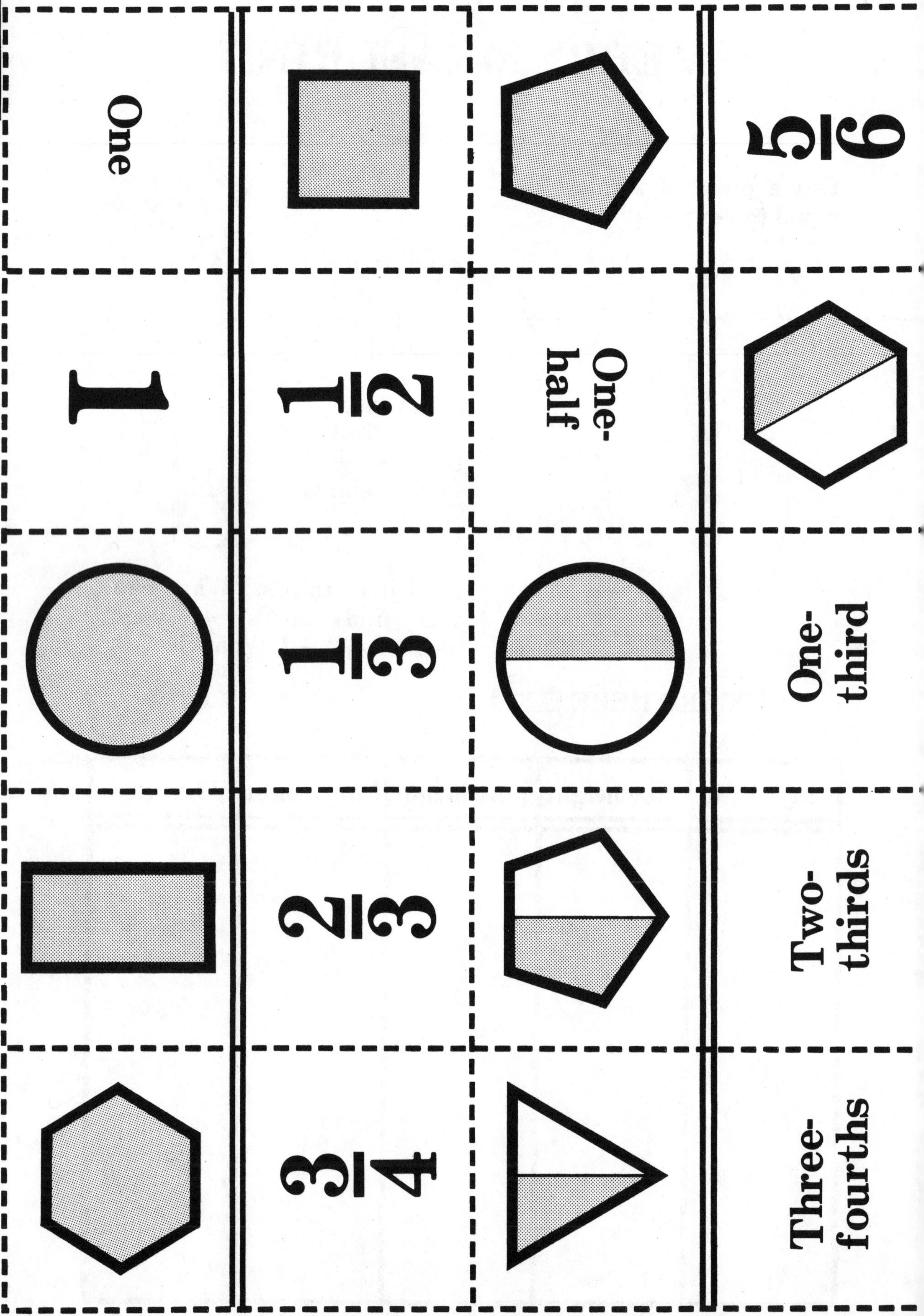

One

1

$\dfrac{1}{2}$

One-half

$\dfrac{5}{6}$

$\dfrac{1}{3}$

One-third

Two-thirds

$\dfrac{2}{3}$

Three-fourths

$\dfrac{3}{4}$

Fraction Dominoes
(Page 2 of three pages)
Back, laminate, and cut along dotted lines.

FRACTIONS

Five-sixths

Fraction Dominoes
(Page 3 of three pages)
Back, laminate, and cut along dotted lines.

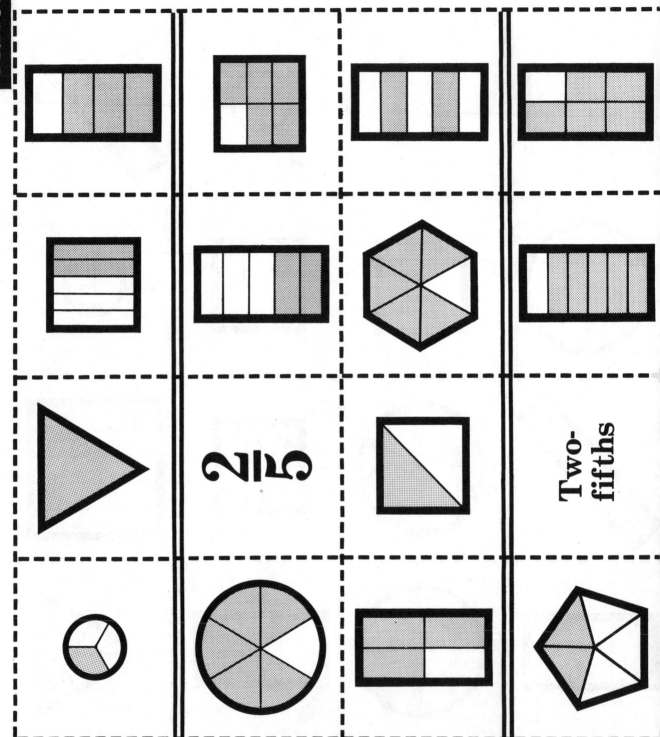

Non-digital Clock Cutout
Back and laminate.
Attach hands with a fastener.

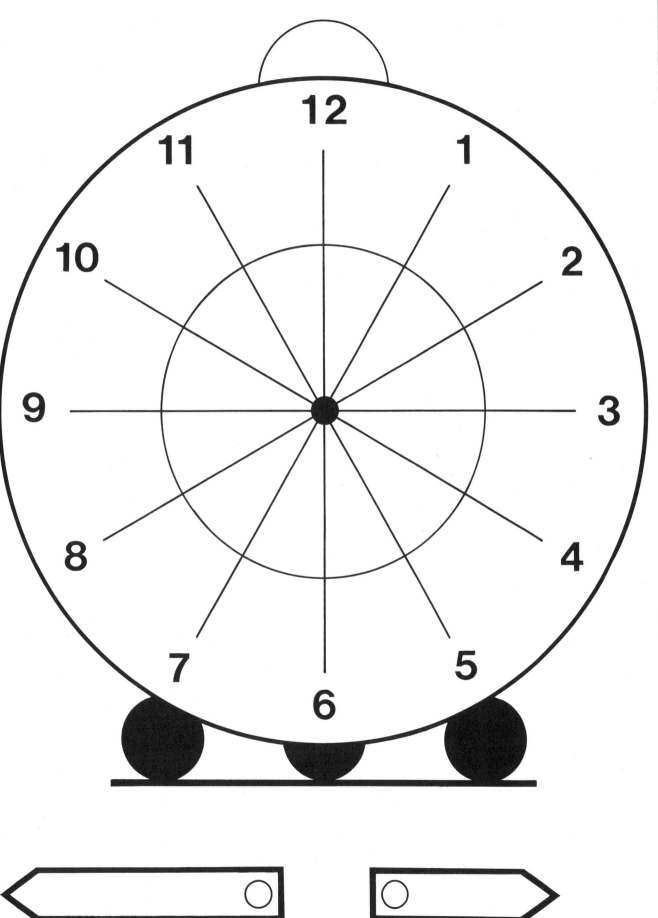

FRACTIONS

Name. . . . .

What time is it?

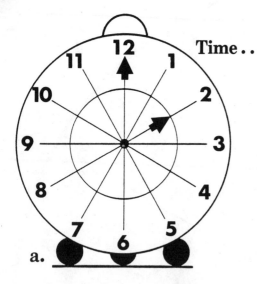 Time. . . . . . .

a.

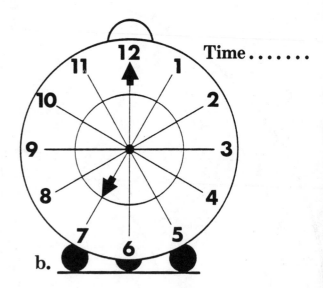 Time. . . . . . .

b.

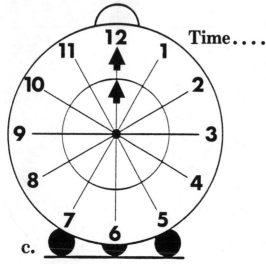 Time. . . . . . .

c.

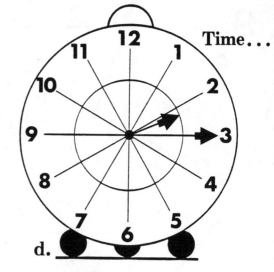 Time. . . .

d.

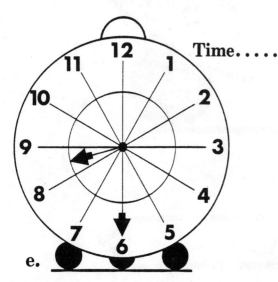 Time. . . . . . .

e.

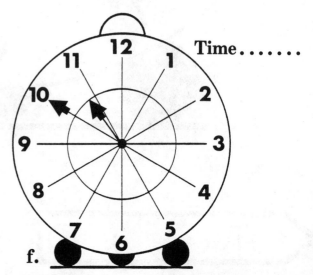 Time. . . . . . .

f.

ame............................

What time is it?

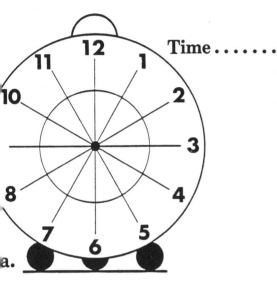

Time.......

a.

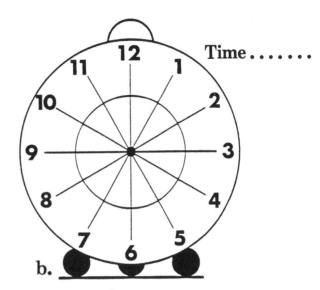

Time.......

b.

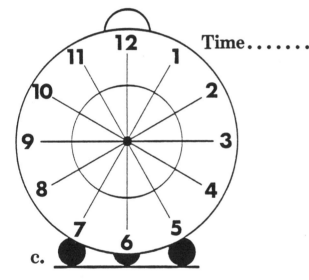

Time.......

c.

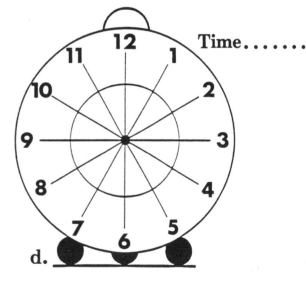

Time.......

d.

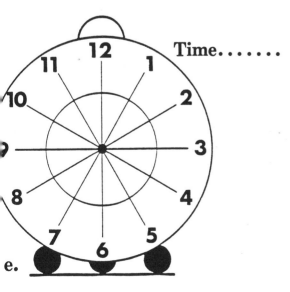

Time.......

e.

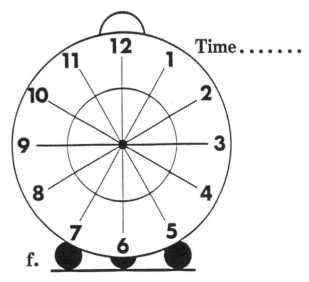

Time.......

f.

Name.............................

Color the shapes. Circle = yellow, triangle = green, square = blue, rectangle = brown, parallelogram = orange, and trapezoid = red.

GEOMETRIC VOCABULARY

Name.............................

How many can you find? S = number of squares, T = number of triangles, R = number of rectangles, TZ = number of trapezoids, and P = number of parallelograms.

S _____ T _____ R _____ TZ _____ P _____

Geo Dominoes
(Page 1 of three pages)
Color, back, and laminate.
Cut along dotted lines.

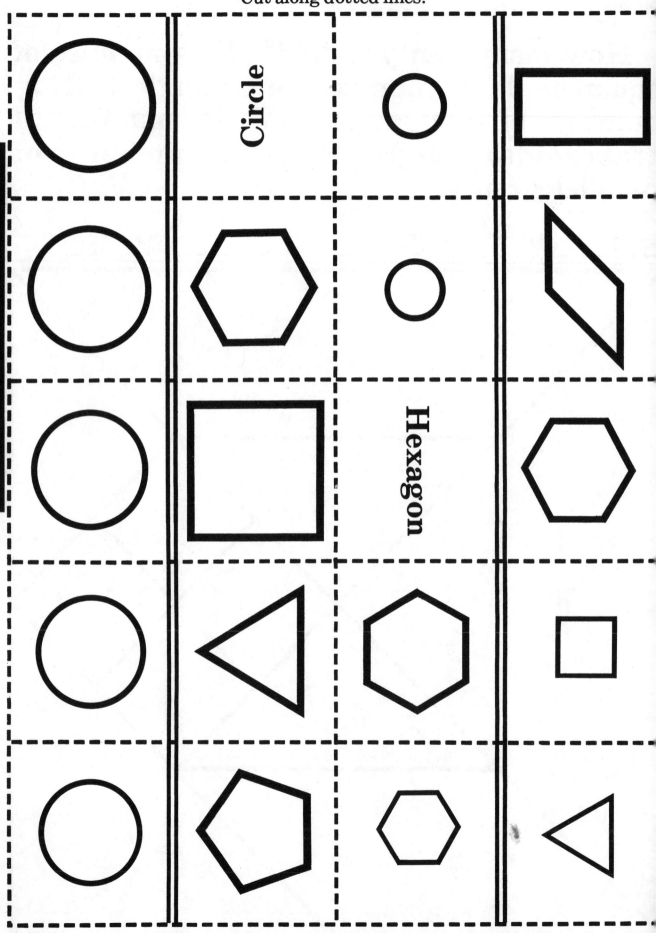

GEOMETRIC VOCABULARY

Circle

Hexagon

Geo Dominoes
(Page 2 of three pages)
Color, back, and laminate.
Cut along dotted lines.

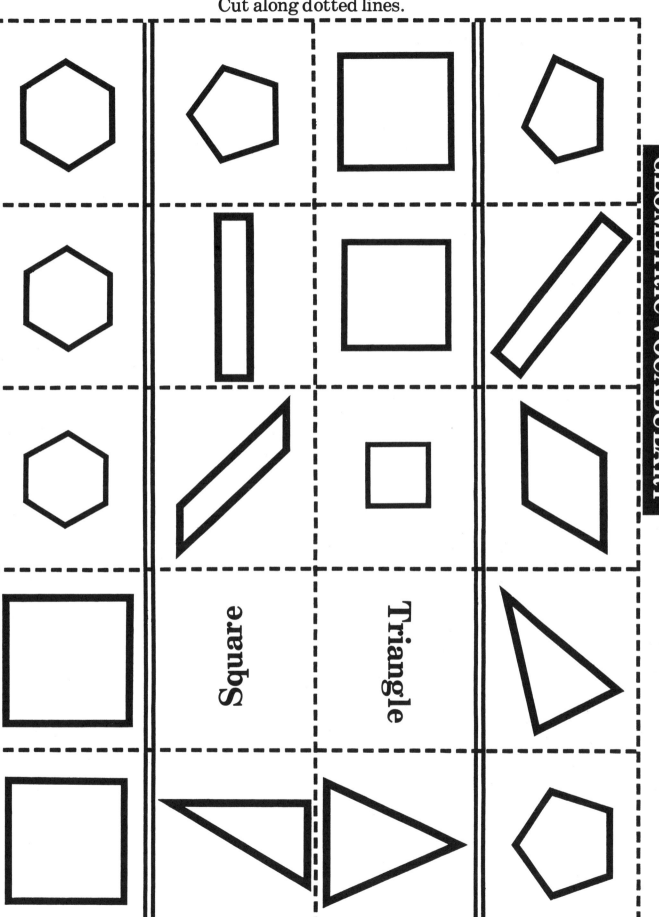

Geo Dominoes
(Page 3 of three pages)
Color, back, and laminate.
Cut along dotted lines.

GEOMETRIC VOCABULARY

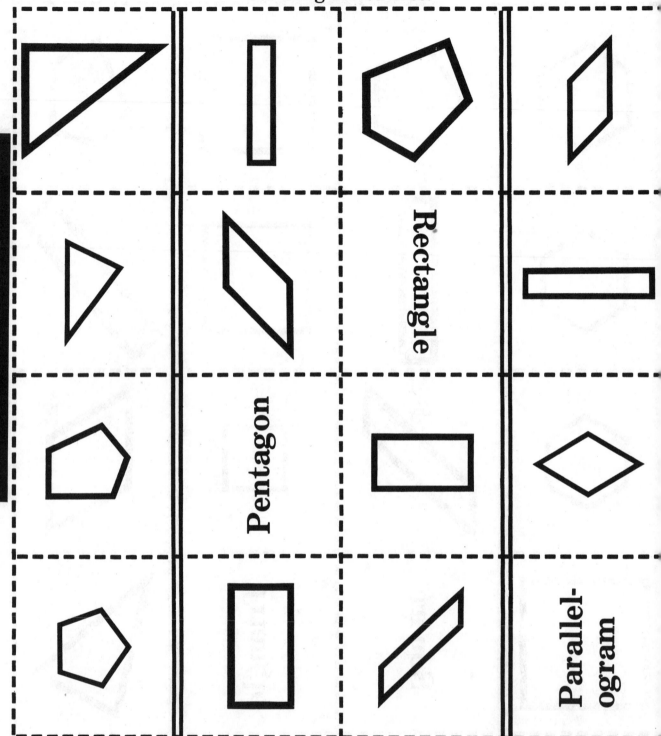

Rectangle

Pentagon

Parallel-ogram

Shape Rummy Cards
(Page 1 of seven pages)
Color shapes, back, and laminate.
Cut along solid lines.

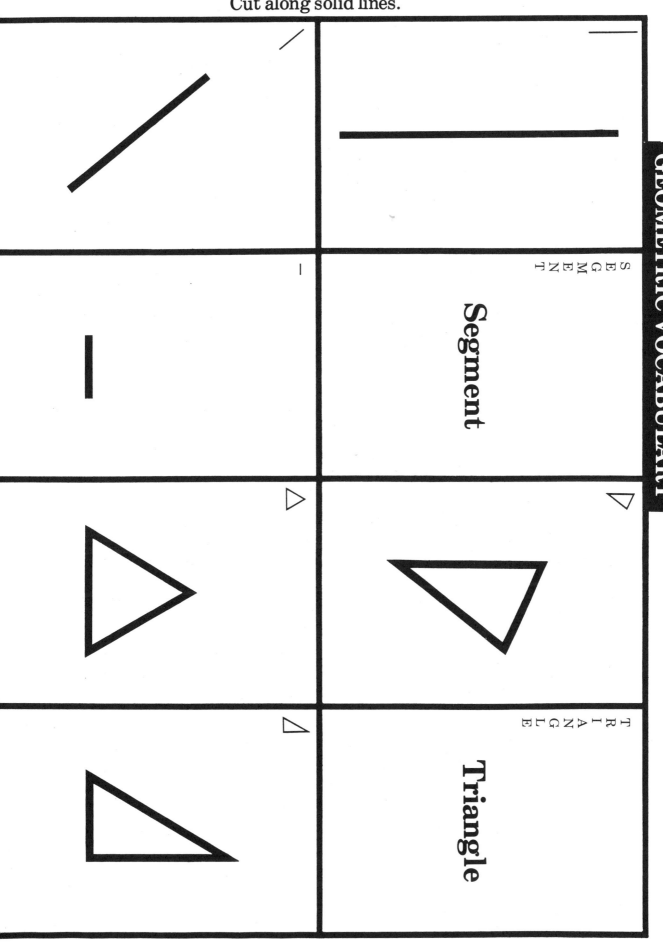

GEOMETRIC VOCABULARY

Segment

S
E
G
M
E
N
T

Triangle

T
R
I
A
N
G
L
E

Shape Rummy Cards
(Page 2 of seven pages)
Color shapes, back, and laminate.
Cut along solid lines.

GEOMETRIC VOCABULARY

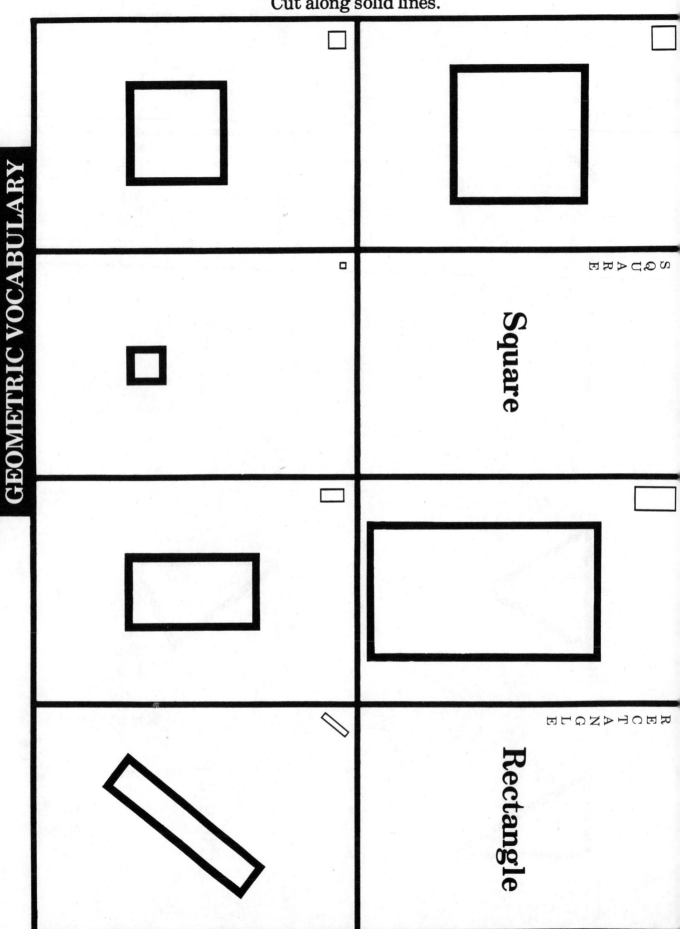

Square

S
Q
U
A
R
E

Rectangle

R
E
C
T
A
N
G
L
E

Shape Rummy Cards
(Page 3 of seven pages)
Color shapes, back, and laminate.
Cut along solid lines.

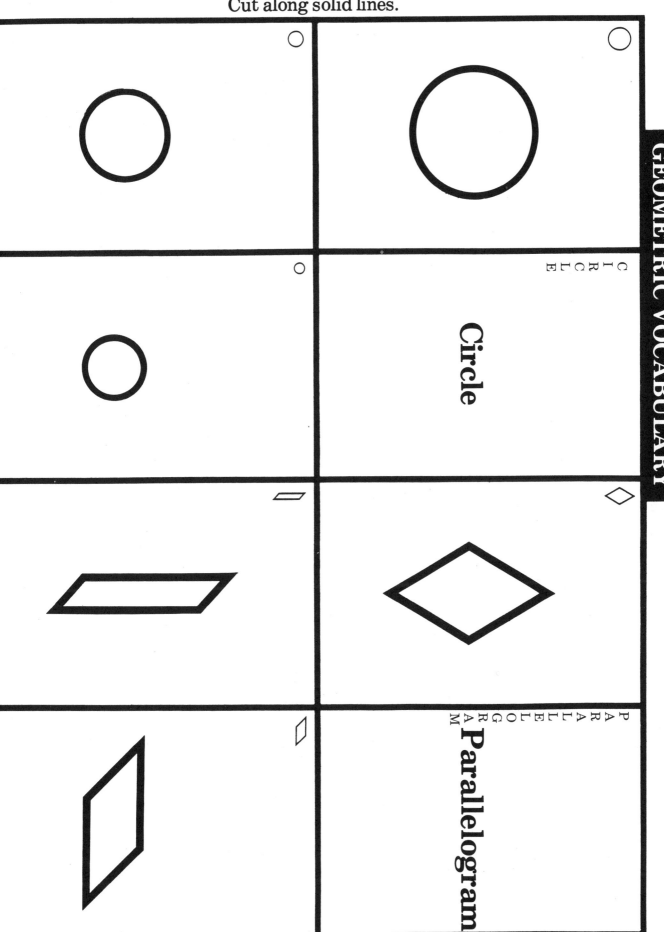

GEOMETRIC VOCABULARY

Circle

CIRCLE

Parallelogram

PARALLELOGRAM

Shape Rummy Cards
(Page 4 of seven pages)
Color shapes, back, and laminate.
Cut along solid lines.

GEOMETRIC VOCABULARY

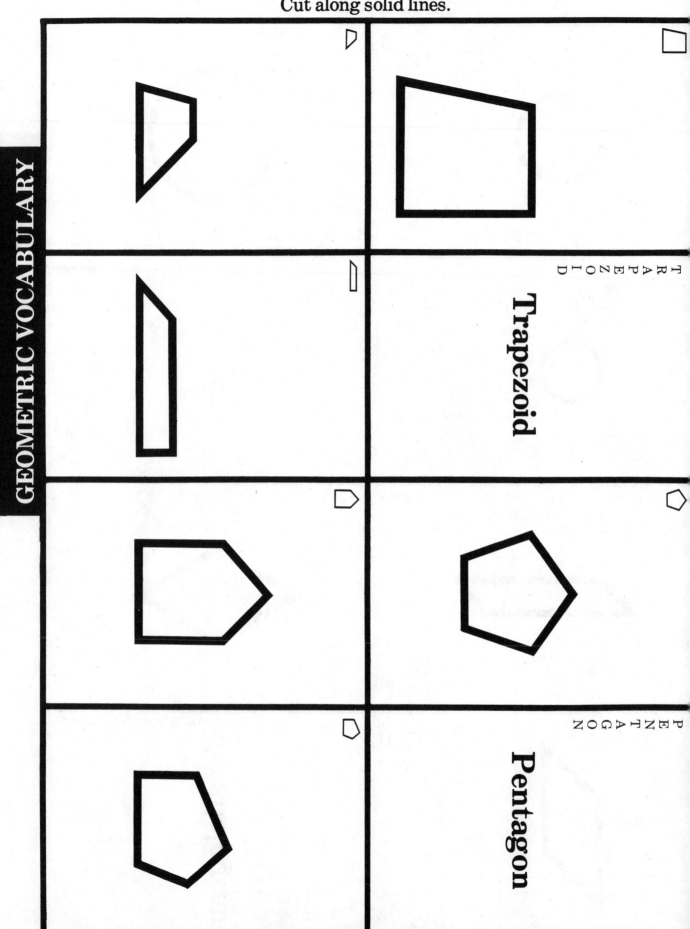

Trapezoid

Pentagon

Shape Rummy Cards
(Page 5 of seven pages)
Color shapes, back, and laminate.
Cut along solid lines.

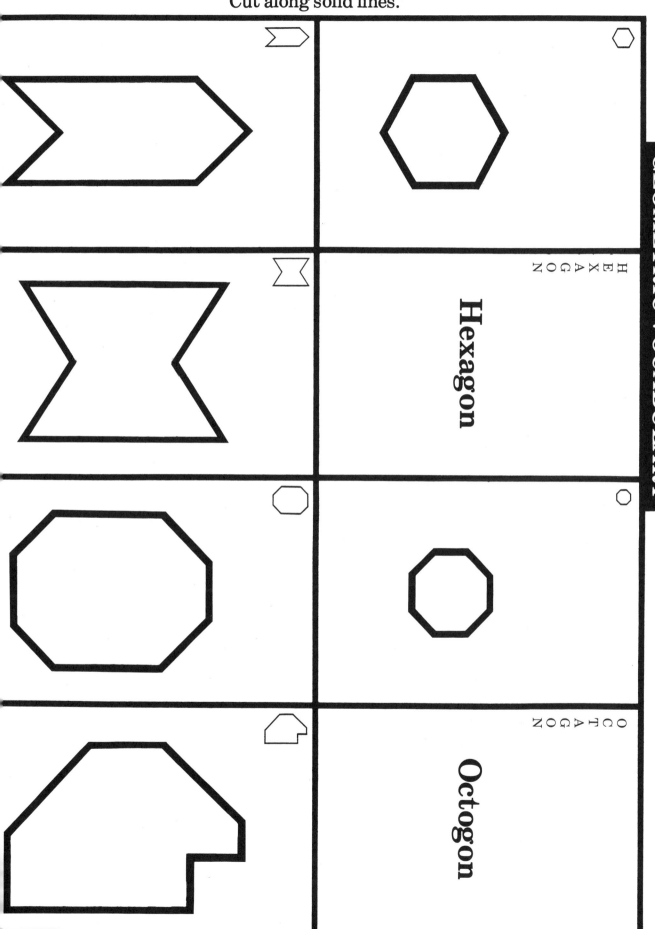

GEOMETRIC VOCABULARY

Hexagon

HEXAGON

Octogon

OCTOGON

Shape Rummy Cards
(Page 6 of seven pages)
Color shapes, back, and laminate.
Cut along solid lines.

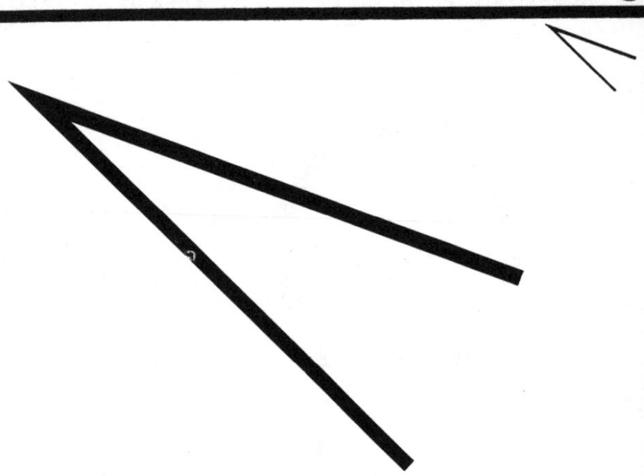

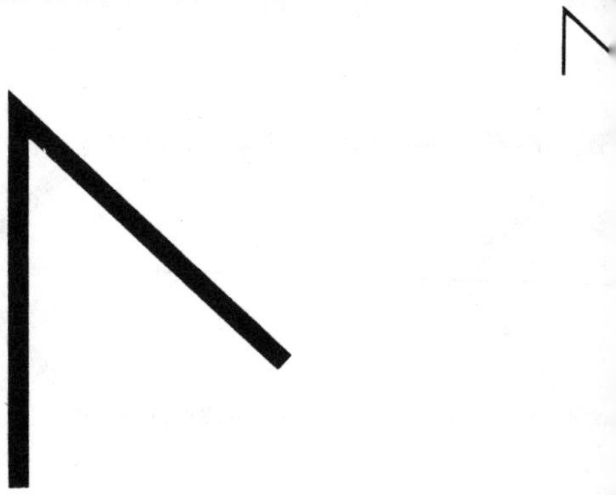

GEOMETRIC VOCABULARY

Acute angle

ACUTE
ANGLE

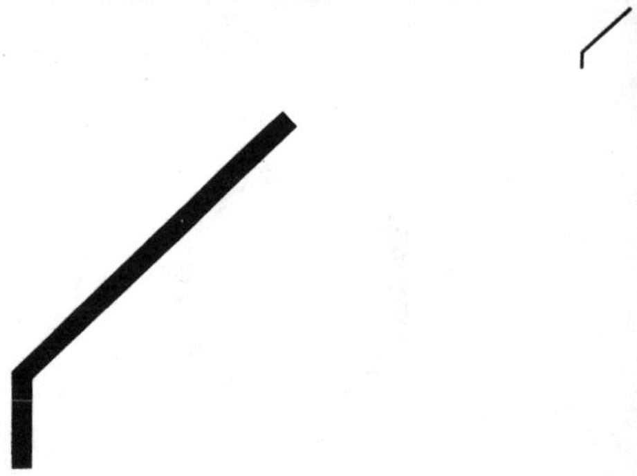

Obtuse angle

OBTUSE
ANGLE

Shape Rummy Cards
(Page 7 of seven pages)
Color shapes, back, and laminate.
Cut along solid lines.

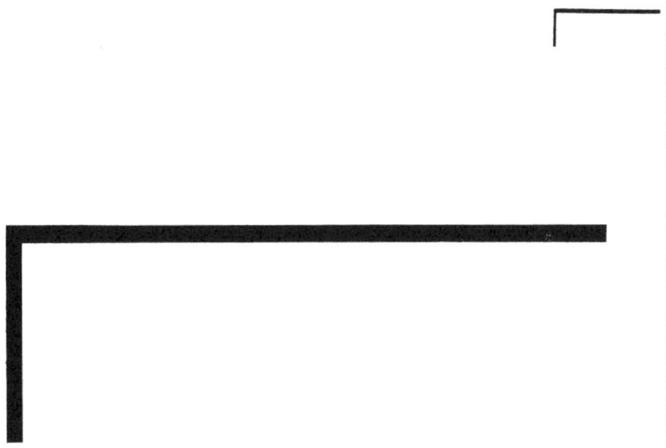

GEOMETRIC VOCABULARY

RIGHT
ANGLE

Right angle

Welcome to Wobble Town. We need workers here. Build the buildings with straws and pipe cleaners. Your teacher wil show you how. When finished, answer the following questions:

1. Which buildings are "wobblies"?

2. Which buildings are "non-wobblies"?

3. How can you make the wobblies non-wobblies?

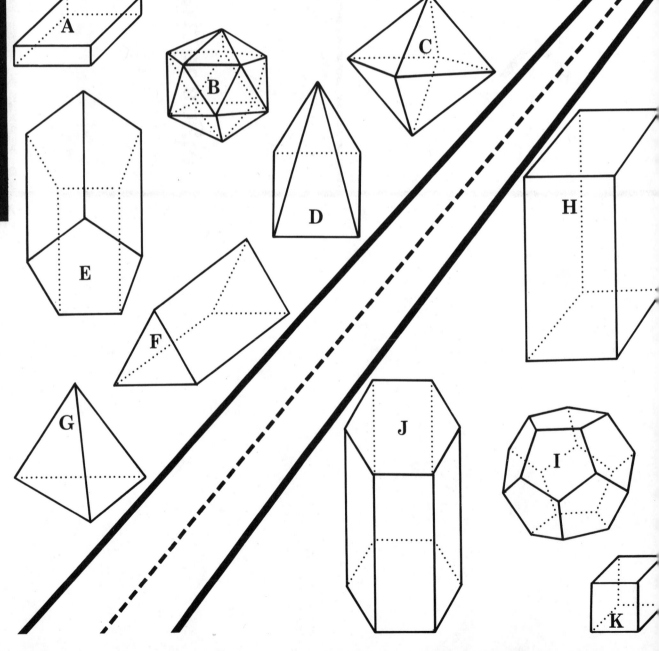

MATH GAMES & ACTIVITIES. COPYRIGHT © 1984

Vertices, Faces, and Edges
(First of three)

Name...

Count the vertices, faces, and edges of each of the figures below and enter your findings in the table. Then complete the table and answer the questions below. Part of the table has already been filled in as a check for your work. V = number of vertices, F = number of faces, and E = number of edges.

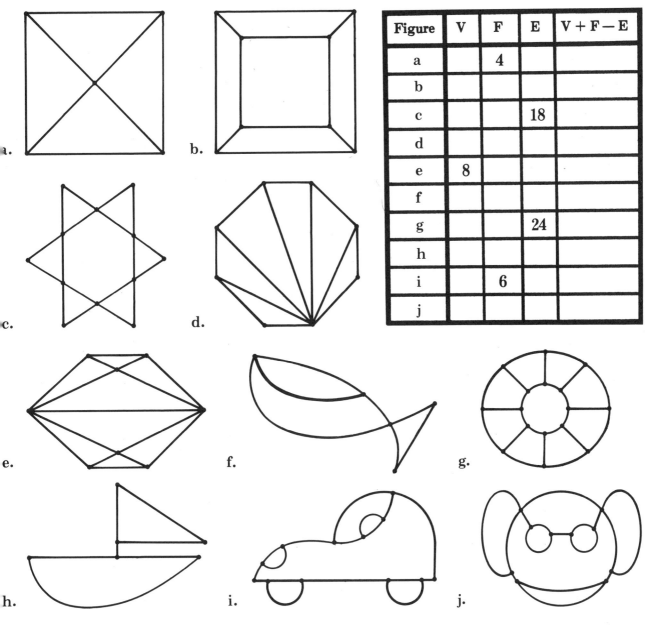

Figure	V	F	E	V + F — E
a		4		
b				
c			18	
d				
e	8			
f				
g			24	
h				
i		6		
j				

. Each of the figures you have been working with is called a network. How would you describe a network?

. If a network had 300 vertices and 250 faces, how many edges do you think it would have?

. If a network had 12,000 vertices and 20,000 edges, how many faces do you think it would have?

. If a network had 31,908,007 faces and 54,732,889 edges, how many vertices do you think it would have?

. How do the vertices, faces, and edges of a network appear to relate to one another?

Vertices, Faces, and Edges
(Second of three)

Name.......................................

Count the vertices, faces, and edges of each of the figures below and enter your finding in the table. Then complete the table and answer the questions below. Part of the table ha already been filled in as a check for your work. V = number of vertices, F = number o faces, and E = number of edges.

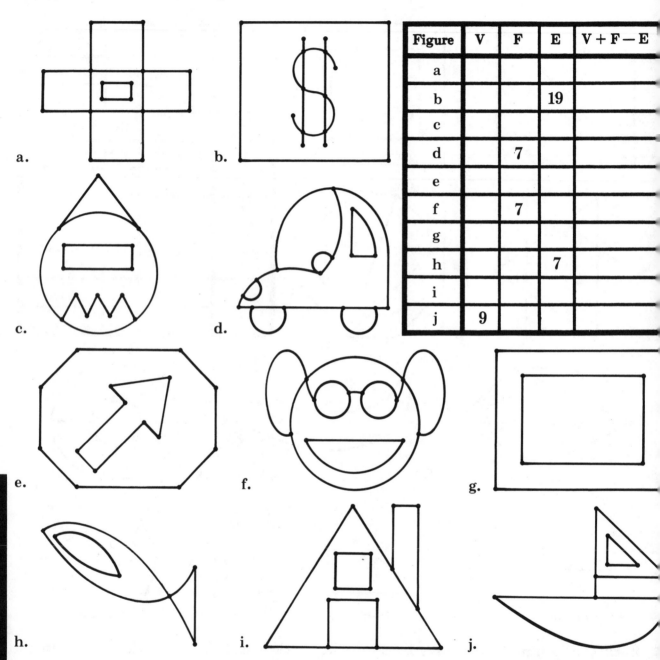

Figure	V	F	E	V + F — E
a				
b			19	
c				
d		7		
e				
f		7		
g				
h			7	
i				
j	9			

a. b. c. d. e. f. g. h. i. j.

Each of the figures you have been working with is how many networks.?

2. How does your answer for the first question relate to V + F — E for each of the figures?

3. Without counting, what is V + F — E for

    a. Each of the small letters of the alphabet?
    b. Each of the capital letters of the alphabet?
    c. Your name if you print it?
    d. Your name if you write it?

VERTICES, FACES, AND EDGES

Vertices, Faces, and Edges
(Third of three)

Name...

Count the vertices, faces, and edges of each of the figures below and enter your findings in the table. Then complete the table and answer the questions below. Part of the table has already been filled in as a check for your work. V = number of vertices, F = number of faces, and E = number of edges.

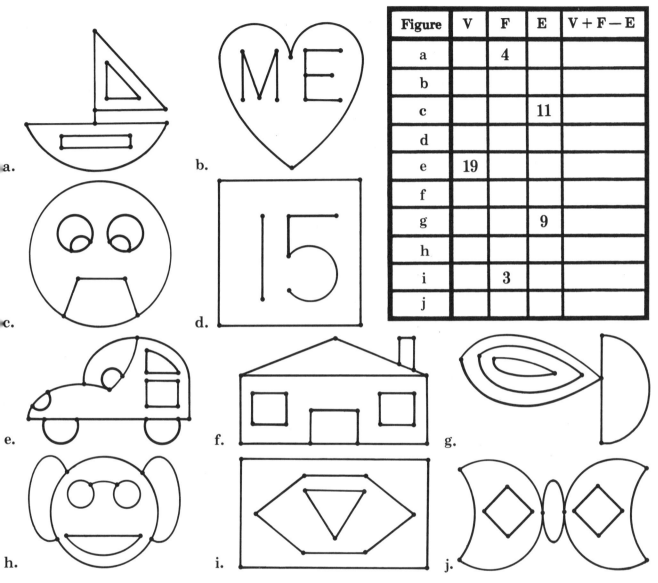

Figure	V	F	E	V + F − E
a		4		
b				
c			11	
d				
e	19			
f				
g			9	
h				
i		3		
j				

a.

b.

c.

d.

e.

f.

g.

h.

i.

j.

1. Each of the figures you have been working with is how many networks?

2. What is an easy way to find V + F − E for a figure?

3. Without counting, what is V + F − E for

a.

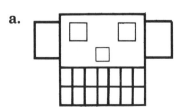

b. A picture of yourself?

VERTICES, FACES, AND EDGES

Mirror Symmetry
(Mirror symmetric figures)

Name.......................................

Six of the figures below are mirror symmetric. Find all six of them. Use a mirror to help you.

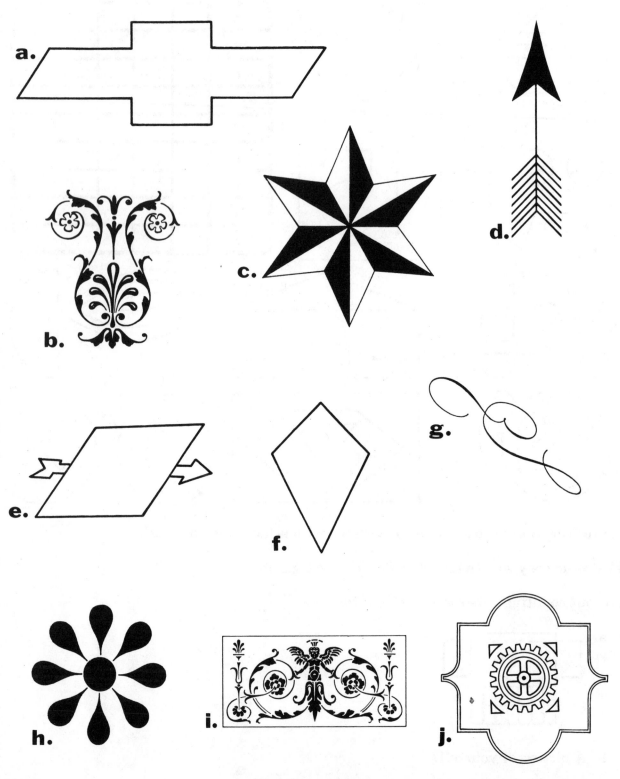

a.

b.

c.

d.

e.

f.

g.

h.

i.

j.

Mirror Symmetry
(Lines of symmetry)

Name...

Some of the figures below are mirror symmetric. Find all the mirror symmetric ones and draw in their lines of symmetry. Use a mirror to help you. There are 30 lines of symmetry in all.

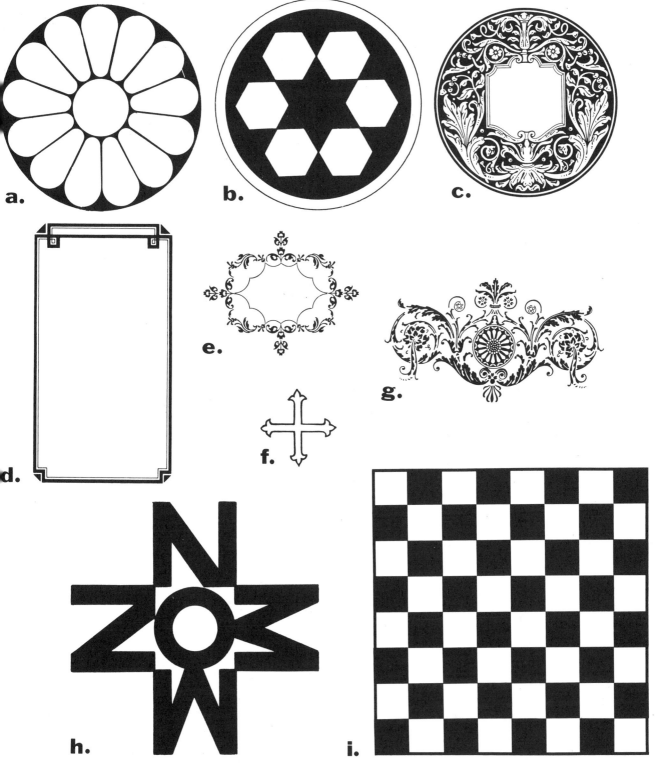

a.

b.

c.

d.

e.

f.

g.

h.

i.

Eye

Name........................

Complete the pattern.

MATH GAMES & ACTIVITIES. COPYRIGHT © 1984

God's Eye

Name..........................

Complete the pattern.

SYMMETRY

Frisbee

Name..........................

Complete the pattern.

Basket Ball

Name...........................

Complete the pattern.

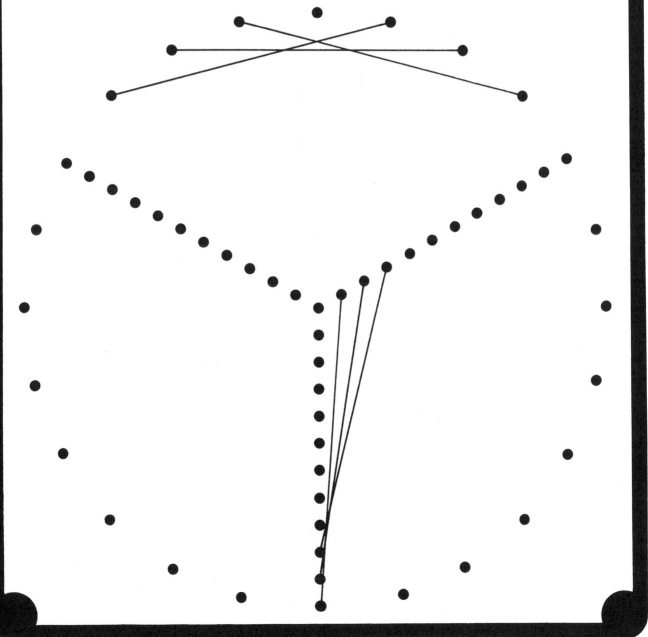

LINEAR MEASUREMENT

FIND THE CHEESE

FINISH

FINISH

START

START

ME AS A MEASURE

Name............................

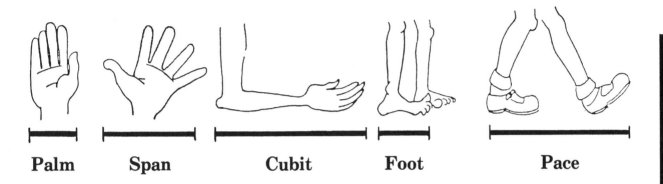

| Palm | Span | Cubit | Foot | Pace |

LINEAR MEASUREMENT

What to measure	Guess	Measurement
Door		
Desk top		
Desk height		
Chalkboard square - - - - High- - - - -		
Book		

Rulers
Back, laminate, and cut out.

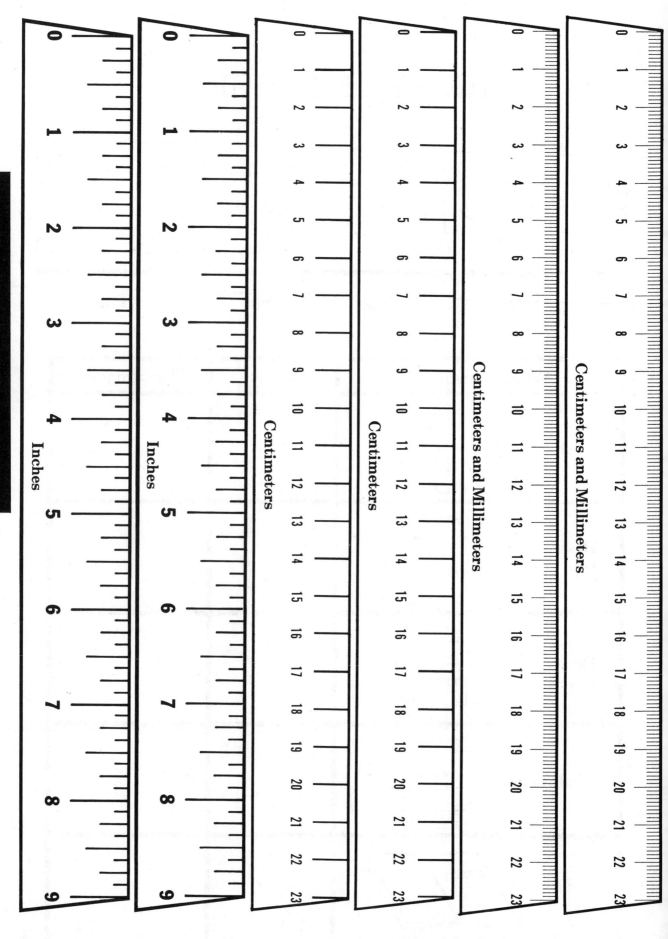

Paths

Name..

Find the shortest path from start to finish. Measure its length in centimeters.

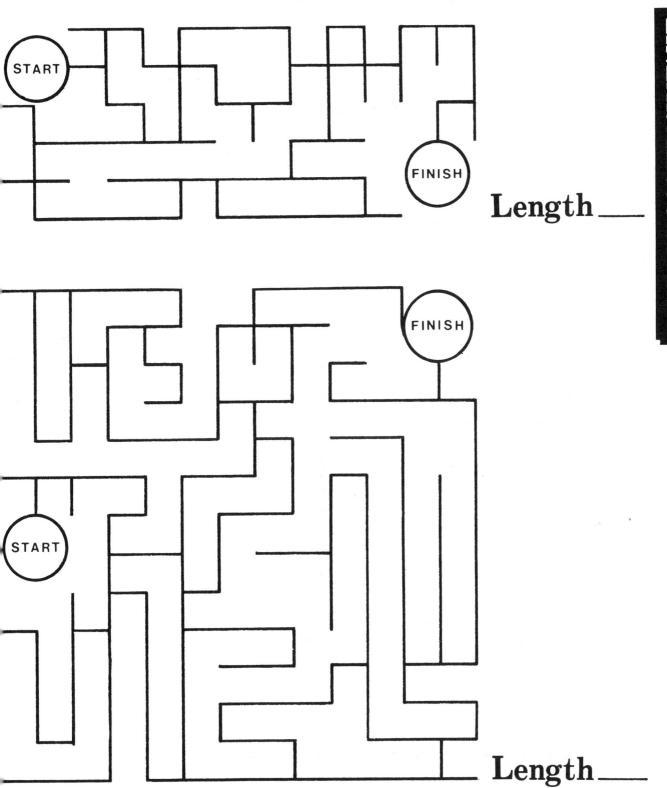

Length ____

Length ____

LINEAR MEASUREMENT

Name...............................

Which is longer? How do you know?

LINEAR MEASUREMENT

1.

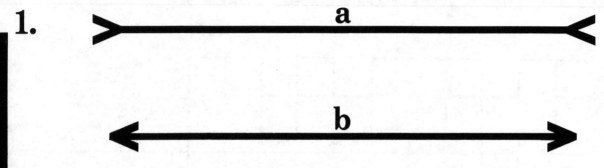

2.

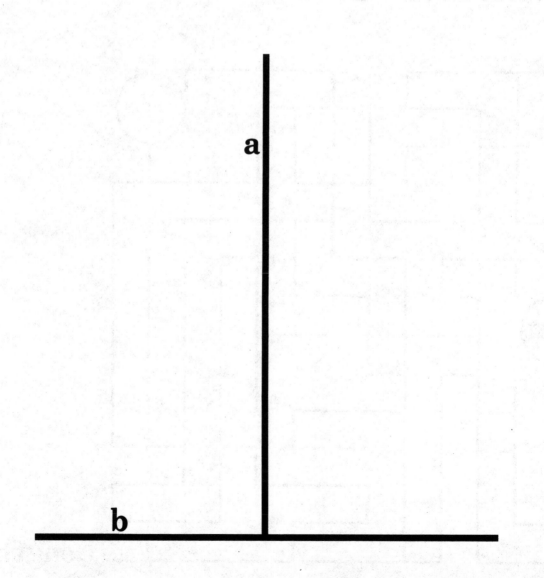

Walk a Crooked Meter Gameboard
Color, back, and laminate.

WALK A CROOKED METER

START

HONEY

FINISH

Ziggy's Home Run Gameboard
(Consumable)

A

D

ZIGGY'S HOME RUN

B

C

Ziggy's Home Run Cards
(Page 1 of four pages)
Back, laminate, and cut along solid lines.

ZIGGY'S HOME RUN	ZIGGY'S HOME RUN	ZIGGY'S HOME RUN
Use the smallest distance in your discard pile.	Use the largest distance in your discard pile.	Use one-half your last measure.
ZIGGY'S HOME RUN	**ZIGGY'S HOME RUN**	**ZIGGY'S HOME RUN**
Use twice your last measure.	Go directly to second base.	Go directly to corner A.
ZIGGY'S HOME RUN	**ZIGGY'S HOME RUN**	**ZIGGY'S HOME RUN**
Go directly to corner B.	Go directly to corner C.	Go directly to corner D.
ZIGGY'S HOME RUN	**ZIGGY'S HOME RUN**	**ZIGGY'S HOME RUN**
1 in.	2 in.	3 in.
ZIGGY'S HOME RUN	**ZIGGY'S HOME RUN**	**ZIGGY'S HOME RUN**
4 in.	5 in.	6 in.

LINEAR MEASUREMENT

Ziggy's Home Run Cards
(Page 2 of four pages)
Back, laminate, and cut along solid lines.

LINEAR MEASUREMENT

ZIGGY'S HOME RUN $1\frac{1}{4}$ in.	ZIGGY'S HOME RUN $1\frac{1}{2}$ in.	ZIGGY'S HOME RUN $1\frac{3}{8}$ in.
ZIGGY'S HOME RUN $2\frac{1}{2}$ in.	ZIGGY'S HOME RUN $2\frac{3}{4}$ in.	ZIGGY'S HOME RUN $2\frac{7}{8}$ in.
ZIGGY'S HOME RUN $3\frac{1}{4}$ in.	ZIGGY'S HOME RUN $3\frac{5}{8}$ in.	ZIGGY'S HOME RUN $3\frac{1}{2}$ in.
ZIGGY'S HOME RUN $4\frac{1}{8}$ in.	ZIGGY'S HOME RUN $4\frac{1}{2}$ in.	ZIGGY'S HOME RUN $4\frac{3}{4}$ in.
ZIGGY'S HOME RUN $5\frac{1}{2}$ in.	ZIGGY'S HOME RUN $5\frac{1}{4}$ in.	ZIGGY'S HOME RUN $6\frac{1}{8}$ in.

Ziggy's Home Run Cards
(Page 3 of four pages)
Back, laminate, and cut along solid lines.

ZIGGY'S HOME RUN	ZIGGY'S HOME RUN	ZIGGY'S HOME RUN
1 cm	**2 cm**	**3 cm**
ZIGGY'S HOME RUN	ZIGGY'S HOME RUN	ZIGGY'S HOME RUN
4 cm	**5 cm**	**6 cm**
ZIGGY'S HOME RUN	ZIGGY'S HOME RUN	ZIGGY'S HOME RUN
7 cm	**8 cm**	**9 cm**
ZIGGY'S HOME RUN	ZIGGY'S HOME RUN	ZIGGY'S HOME RUN
10 cm	**11 cm**	**12 cm**
ZIGGY'S HOME RUN	ZIGGY'S HOME RUN	ZIGGY'S HOME RUN
13 cm	**14 cm**	**15 cm**

LINEAR MEASUREMENT

Ziggy's Home Run Cards
(Page 4 of four pages)
Back, laminate, and cut along solid lines.

ZIGGY'S HOME RUN	ZIGGY'S HOME RUN	ZIGGY'S HOME RUN
1.2 cm	**2.6 cm**	**3.4 cm**
ZIGGY'S HOME RUN	ZIGGY'S HOME RUN	ZIGGY'S HOME RUN
4.5 cm	**5.1 cm**	**6.0 cm**
ZIGGY'S HOME RUN	ZIGGY'S HOME RUN	ZIGGY'S HOME RUN
7.7 cm	**8.9 cm**	**9.8 cm**
ZIGGY'S HOME RUN	ZIGGY'S HOME RUN	ZIGGY'S HOME RUN
10.3 cm	**11.5 cm**	**12.9 cm**
ZIGGY'S HOME RUN	ZIGGY'S HOME RUN	ZIGGY'S HOME RUN
13.1 cm	**14.5 cm**	**15.2 cm**

Metric Concentration Cards
(Page 1 of two pages)
Back, laminate, and cut along solid lines.

METRIC CONCENTRATION	METRIC CONCENTRATION
1 m	**100 cm**
METRIC CONCENTRATION	METRIC CONCENTRATION
	10 cm
METRIC CONCENTRATION	METRIC CONCENTRATION
1 cm	**10 mm**
METRIC CONCENTRATION	METRIC CONCENTRATION
	1 mL

LINEAR MEASUREMENT

Metric Concentration Cards
(Page 2 of two pages)
Back, laminate, and cut along solid lines.

LINEAR MEASUREMENT

METRIC CONCENTRATION	METRIC CONCENTRATION
1 km	**1000 m**
METRIC CONCENTRATION	METRIC CONCENTRATION
1 L	**1000 mL**
METRIC CONCENTRATION	METRIC CONCENTRATION
1 kg	**1000 g**
METRIC CONCENTRATION	METRIC CONCENTRATION
1 g	**1000 mg**

Grid Paper
(2-centimeter)

AREA

Grid Paper
(1-centimeter)

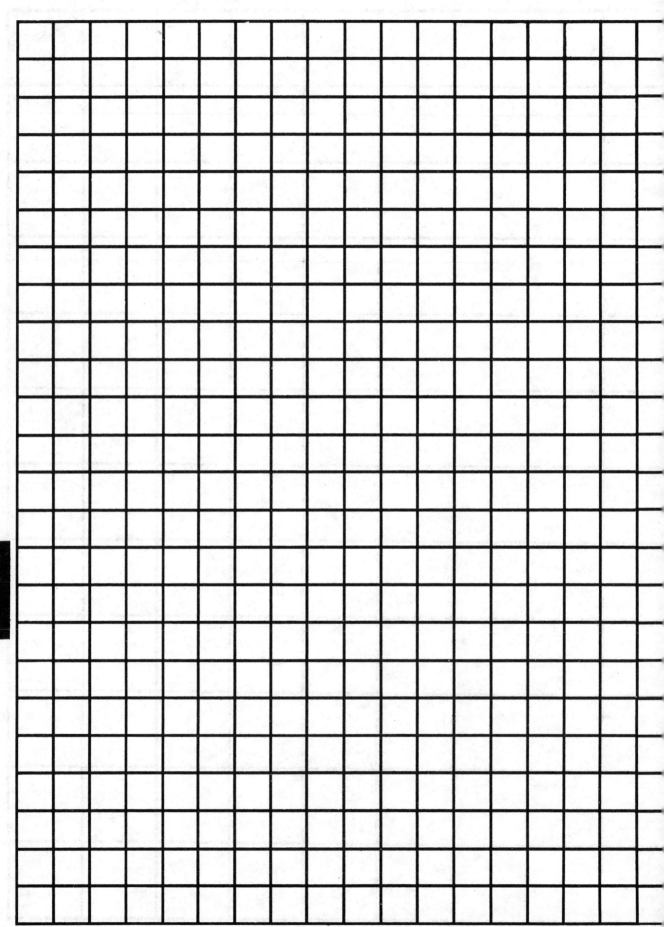

AREA

Tangram Pieces
(Two sets)
Back, laminate, and cut along solid lines.

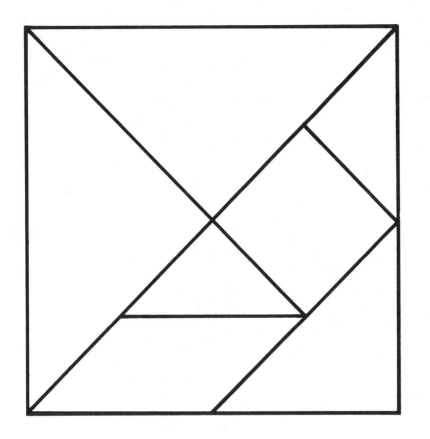

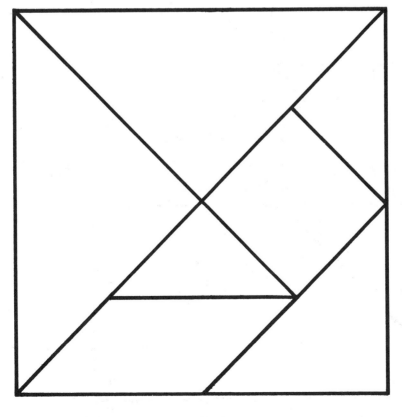

Tangram Task Cards
(Bird and house)
Color, back, and laminate.

Cover me
with Tangr-
am Pieces.

Cover me
with Tangr-
am Pieces.

Tangram Task Cards
(Boat and fish)
Color, back, and laminate.

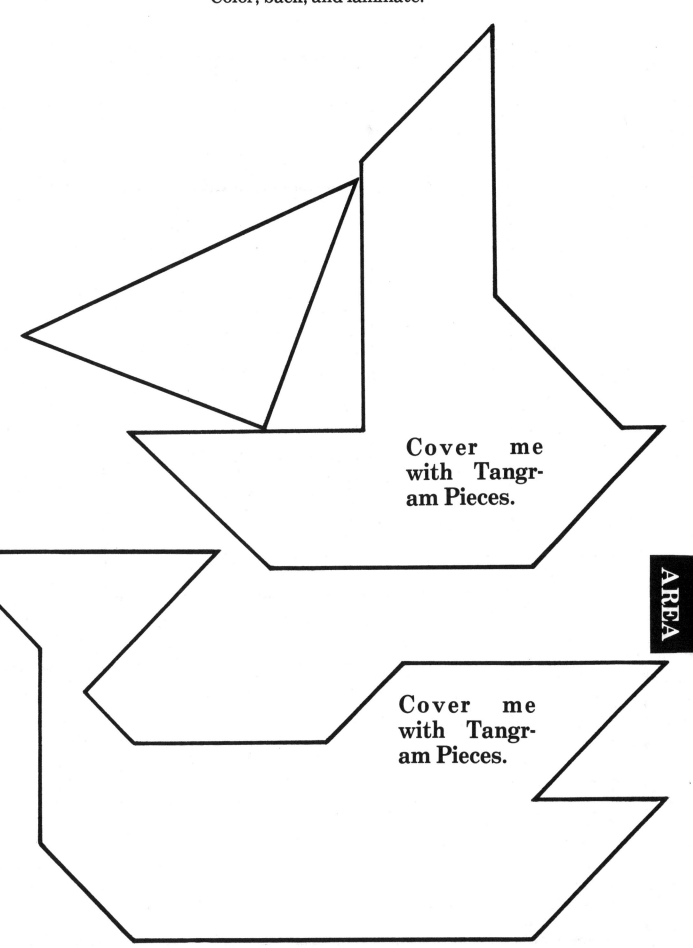

Cover me with Tangram Pieces.

Cover me with Tangram Pieces.

AREA

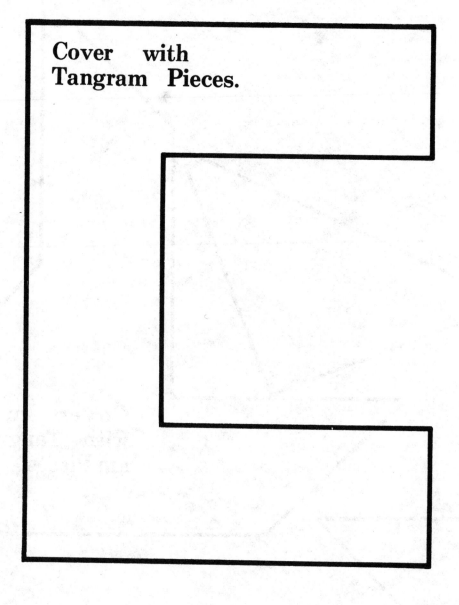

Cover   with
Tangram   Pieces.

AREA

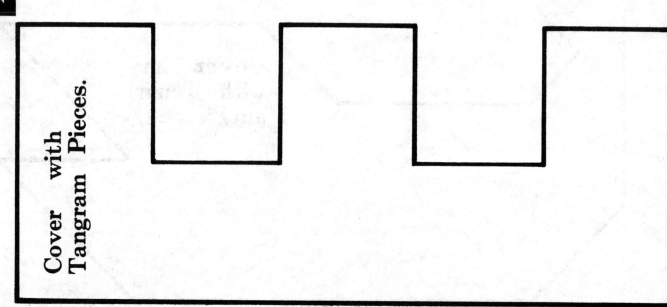

Cover   with
Tangram Pieces.

Tangram Task Cards
(Bridge and kettle)
Color, back, and laminate.

Make the bridge with Tangram Pieces.

Make the kettle with Tangram Pieces.

AREA

Tangram Task Cards
(Cat and candle)
Color, back, and laminate.

Make the cat with Tangram Pieces.

Make the candle with Tangram Pieces.

AREA

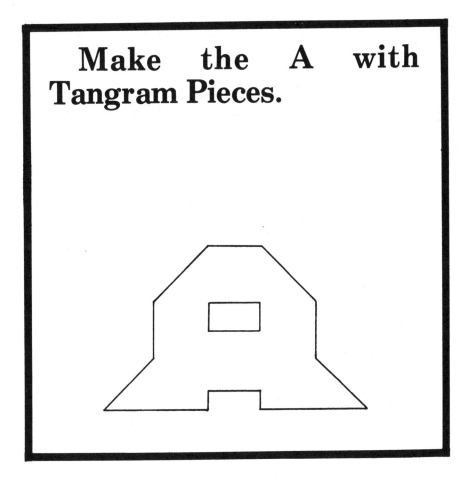

Make the A with Tangram Pieces.

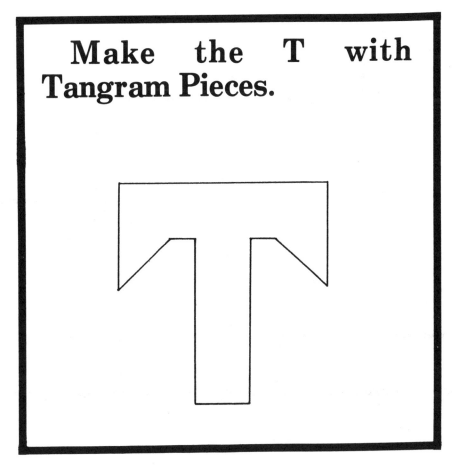

Make the T with Tangram Pieces.

AREA

Tangram Task Card
(Comparing)
Color, back, and laminate.

Is one bigger than the other? Find out with Tangram Pieces.

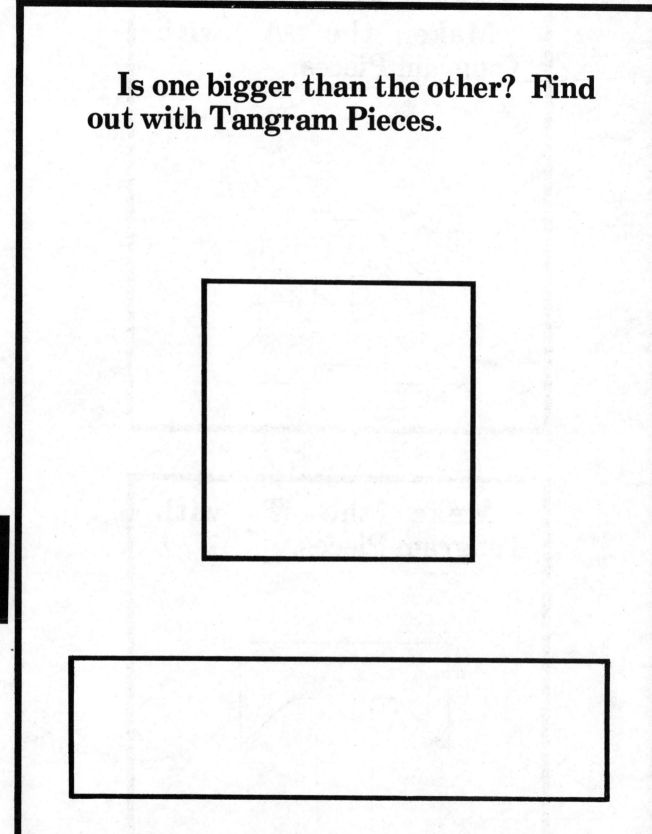

AREA

Which two have the same area?
Find out with Tangram Pieces.

AREA

Milliliter, 10-Milliliter, and 100-Milliliter Cutouts

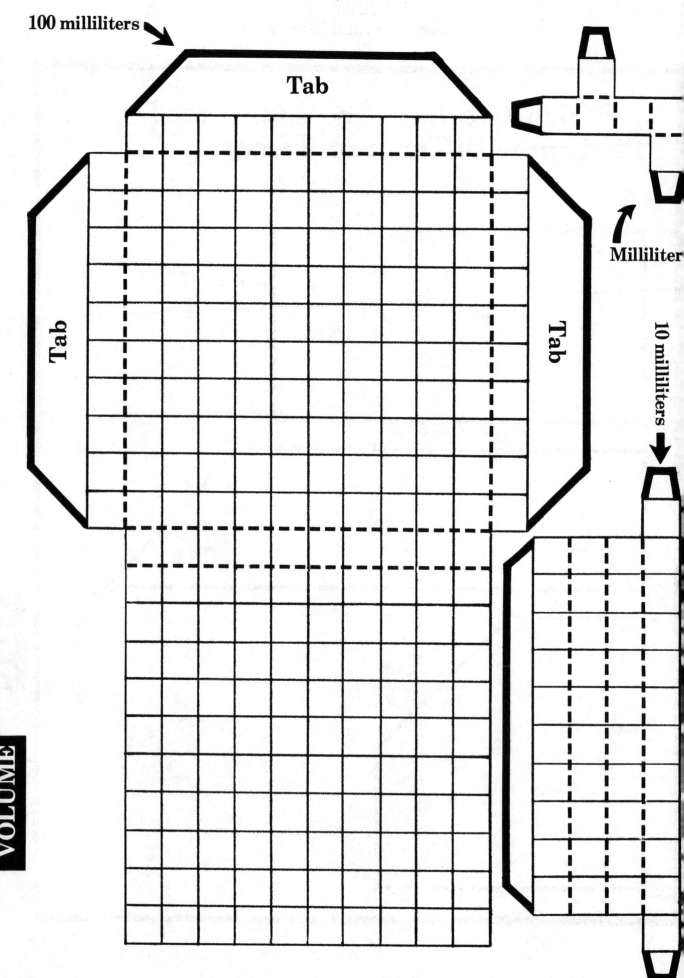

100 milliliters

Tab

Tab

Tab

Milliliter

10 milliliters

VOLUME

Liter Cutout
(Piece 1 of two pieces)

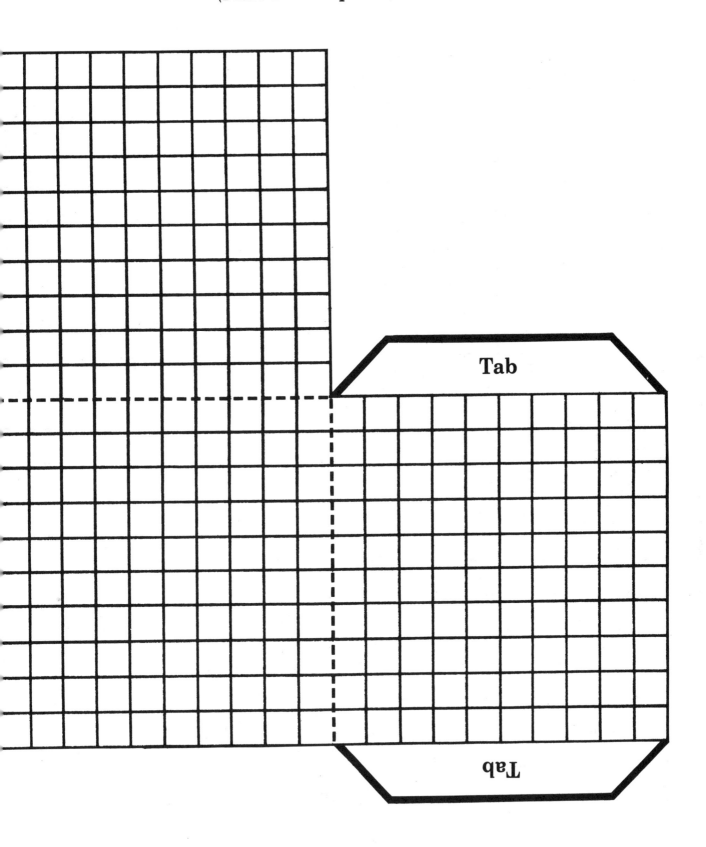

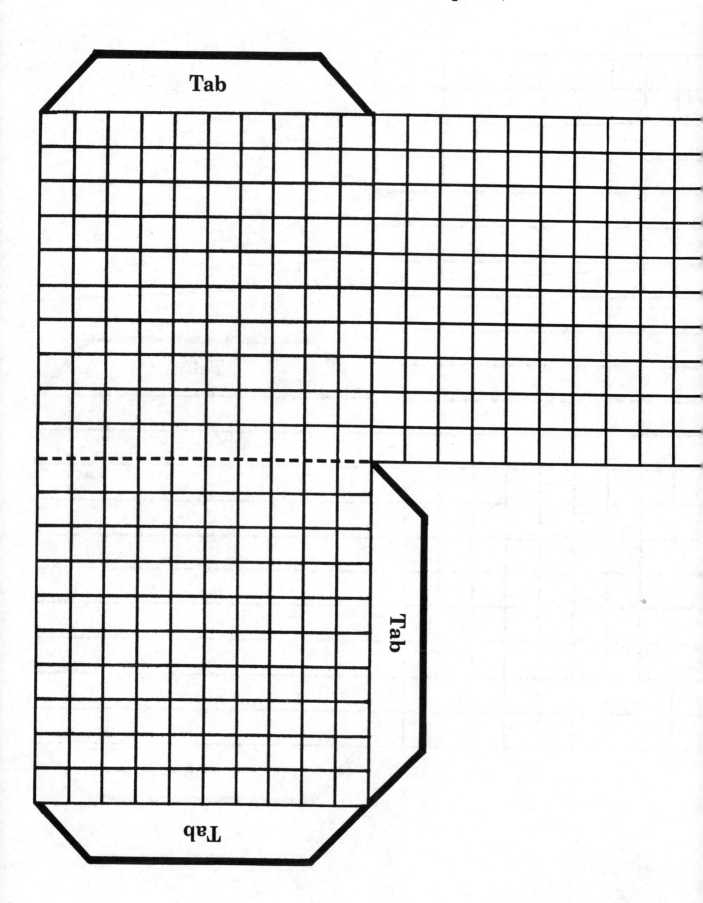

INDEX*

*The page numbers in the index refer to the page numbers of the duplicating designs.

Bibliography

Baratta-Lorton, M. **Mathematics Their Way,** Addison-Wesley Publishing Company, Inc., 1976.

Baratta-Lorton, M. **Workjobs II,** Addison-Wesley Publishing Company, Inc., 1979.

Broadbent, F.W. "'Contig': a game to practice and sharpen skills and facts in the four fundamental operations," **Games and Puzzles for Elementary and Middle School Mathematics,** National Council of Teachers of Mathematics, 1975.

Bruni, J.V. and H. Silverman. "Making and using board games," **The Arithmetic Teacher,** 22: 173-179, 1975.

Burns, M. "Ideas," **The Arithmetic Teacher,** 21: 506-518, 1974.

Court, N.A. **Mathematics in Fun and in Earnest,** The New American Library of World Literature, Inc., 1935.

"Cover Ollie Octopus," **Rhombus,** A journal of the Mathematical Association of Western Australia, 7: 14-19, 1979.

Cruikshank, D.E. "Sorting, classifying, & logic," **The Arithmetic Teacher,** 21: 588-598, 1974.

Di Spigno, J. "Division isn't that hard," **The Arithmetic Teacher,** 18: 373-377, 1971.

Dumas, E. and C.W. Schminke. **Math Activities for Child Involvement,** Allyn and Bacon, Inc., 1977.

Hutchings, B. "Low-Stress Algorithms," **Measurement in School Mathematics,** 1976 yearbook of the National Council of Teachers of Mathematics.

Judd, W. **Math Mat Activities,** Creative Publications, Inc., 1974.

Knaupp, J. and G. Knamiller. **Open Math,** Arizona State University, 1974.

Krause, M. "Creative classroom: Metric Shake," **Teacher**, Oct., 1978.

Latta, R. "Creative classroom: Number-go-Round," **Teacher**, Mar., 1979.

McLean, P. and B. Sternberg. **People Piece Primer**, Activity Resources Company, Inc., 1975.

Midwest Publications Company, Inc., P.O. Box 129, Troy, Michigan, 48048.

Nelson, R.S. "Variations on rummy," **The Arithmetic Teacher**, 25: 40-41, 1978.

Palmer, R. **Springboards: Ideas for Mathematics**, Thomas Nelson Australia Pty. Ltd., 1979.

Parsons, J. N. **Math.A.Dot**, Levels I, II, and III, Fearon Publishers, Inc., 1975.

Project for the Mathematical Development of Children, Florida State University, Newsletter No. 2, 1975.

Scott Resources, Inc., 1900 East Lincoln, Box 2121 CA, Fort Collins, Colorado 80522.

Seymour, D. **Tangramath**, Creative Publications, Inc., 1971.

Shoecraft, P.J. **Basic Mathematics: A Blueprint for Success**, Addison-Wesley Publishing Company, Inc., 1979.

Wills, H. "Diffy," **Games and Puzzles for Elementary and Middle School Mathematics**, National Council of Teachers of Mathematics, 1975.